THE ALPINE LAKE INCIDENT

BEN ARMSTRONG

ARMSTROB

{armstrob}

First edition

ISBN: 978-1-7342597-4-2

For you, dear reader.

I appreciate you more than I could ever express in words on a page.

CONTENTS

PROLOGUE

JULY 27TH, 2010

Lights and sirens filled the otherwise quiet driveway of 1824 Alpine Way. The normally quiet and serene lake house was now a crime scene—one that would leave a long-lasting impression on the responding officers. Upon their arrival, the only sign of anyone being there was the Henderson's Toyota Camry parked in front of the house. Officers Tom and Rick were the first on scene.

Long past its prime, the aging house had fallen into disrepair. The paint was peeling off from years of the sun beating down on it, and the porch looked as if it could collapse at any moment. The lawn was overgrown with weeds, making the path leading up to the house nearly impossible to see. It looked as if no one had actually been there in a very long time.

"Ugh, why would anyone want to live in this place?" asked Tom, getting out of the passenger seat of the squad car.

"What do you mean?"

"You don't know the history of Alpine Lake, do you?

There's been a lot of 'weird' happenings out here over the years. I remember one in particular—"

"Holy shit," said Rick, pointing towards the house.

As Tom turned, he saw a girl standing on the front porch, her hair and face drenched in blood. So much so, it ran down her arms, pooling at her fingertips. The two officers cautiously approached her, still unsure of what exactly they were dealing with. She had an old leather journal clenched in her hand, her utterly blank stare unnerving them both. Rick waved his hand in front of her, trying to get a reaction—but there was nothing.

"Hello?" asked Rick. "Little lady, are you okay? What happened here?" The girl continued to stare through the two of them, appearing to be unaware of their existence.

"Rick, stay with the girl. I'm going to take a look inside." He walked up to the porch and into the house. Not more than a minute later, he ran back outside, his face void of any color. "Jesus Christ, we're going to need backup here. I'm going to make a few calls."

"What'd you find?"

Tom hesitated for a moment. "I—I don't think you'd believe me. Stay with the girl, I'm going to get backup. It's going to be a long night."

FOREWORD

The following is from the writings of Chloe Henderson, Jessica Ambrose, and Dr. Lily Burke. Nothing has been altered from the originals. The audio recordings have been transcribed word for word.

CHLOE

2023 (PRESENT DAY)

Hello there, I hope you're doing well. As for me, I've been staring at a blank Word document on and off for the past two weeks, each time watching the blinking cursor as it mocks me for not being able to come up with how to start this. I was beginning to think I wouldn't be able to write anything. My therapist swears it will help me, however, so I'm determined to do it.

Oh, where are my manners? My name is Chloe Henderson. I'm twenty-nine as of writing this, and I currently live in Sioux Falls, South Dakota. It's just me and my two cats right now. After what happened, which is the intended subject of this book, or journal, or whatever you want to call it, I haven't really been able to focus on any sort of relationship, hence the two cats and heavy journaling. Again, my therapist thinks getting this out there will help. I hope it does.

I've been seeing my therapist for a number of years now, although there have been times when I will go months without seeing her in an attempt to work on things myself. In the end, I always give in and start to see her again on a regular basis.

Her name is Lily, and I really like her. She's been there for me through some of the toughest parts of my life. She was there just a short week after *it* happened, and I don't think I could ever repay her for all her help—monetary compensation aside.

As I sit here writing this intro, I'm trying to decide where to start the story. Despite *it* happening nearly thirteen years ago, the events leading up to that warm summer night in July still haunt my dreams on a regular basis. I've read about several cases where traumatic events happen to people, particularly children, and their minds push those memories away to protect them from the trauma. For whatever reason, mine decided to keep them at the forefront of my memories—lucky me.

Anyway, I should start writing this story before I decide it's not worth it. My therapist wants me to have at least one chapter written by the time I see her again, which is in two weeks. I think I know where to start. I want to apologize in advance if certain parts of this don't make complete sense, or if there are gaps in certain places.. While these memories feel like they just happened to me yesterday, there are still parts of the events my brain refuses to let me remember—whatever that means.

CHAPTER ONE

1

In the Summer of 2010, my family and I lived in a rundown apartment just south of Main Street in Fargo, North Dakota. With four of us living under the small roof, two bedrooms and one bathroom made for a cramped space. Could it have been worse? Yes, but it could also have been much better. I said it was rundown, but in reality it was barely livable. Our apartment was on the third floor, which happened to be the top, and one of the few 'nice' things about it was it being a corner unit.

The morning of June 25th started like any other. My mom made my sister and me breakfast, eggs and toast being the norm, and afterwards we would wait for the bus to school. My dad worked at the Inforum, a local newspaper, so he was always up and out of the house by six a.m. Meanwhile, my mom would stay at home and do her best to keep the apartment in working order. She made the best of our not so great situation, and we were all appreciative of it.

"Good morning, Mom," I said, coming into the kitchen from my bedroom. She was already hard at work preparing breakfast for us. "Smells good. I'm starving."

She turned to me and smiled. "Good morning, dear. Is your sister up?"

I nodded. "Yup, she's getting dressed. I asked if she wanted any help with her hair, but you know her. 'I can do it myself!'"

She laughed. "Yes, that sure sounds like her. Breakfast is almost ready."

A moment later my sister comes running into the kitchen with her hair in disarray. Charlotte, who just recently turned ten, was by far the feistier of us girls. She's a great sister, don't get me wrong, she can just be a little too stubborn at times. Mom says it's just the age she's at and that I went through the same thing.

"See? I didn't need any help with my hair." She plopped herself down in the chair next to me.

I did my best not to chuckle. "Looks great, Char."

She hated being called Char, but I was the only one she actually let call her that. My mom brought over two plates and put them down in front of us. Her cooking was always great, despite the pans and the stove being older than both my sister and me.

"Eat up," she said. "It's important to have a good breakfast to get you going in the morning." Mom said this same thing nearly every morning. Some might find it annoying, but I personally found it to be endearing. She loved us so much.

Char let out a belch and giggled. "This is super good, Mom."

"What do you say, Charlotte?" Mom looked at her with feigned disappointment, suppressing a laugh herself.

"Excuse me!" She laughed again. Char never was one to shy away from eating food at an absurd speed when it was in front of her. I took my time, but enjoyed the food just as much. After we had both cleaned our plates, we kissed our mom goodbye and left to wait for the bus outside.

2

The final bell of the school day rang at 3:30, and I walked outside to wait for Char at our normal meeting spot next to the bus pickup. School had been good, as it almost always tended to be for me. Unlike many kids my age, I thoroughly enjoyed it. Not because of seeing my friends or anything—I didn't have many of those. I just loved to learn things. Most of all, I loved to read, which made English my favorite subject.

We're currently reading *1984 by George Orwell*. It's a terrifying look at what could very possibly be our reality in the not too distant future. I think I'm the only one in my class who is bothering to actually read it, but that's okay. Reading is a wonderful escape—one that I can't seem to get enough of. I was just about to read a couple more pages when Char finally made her appearance.

"What took you so long?" I asked, already knowing the answer. She was one of the most distractible people I'd ever met. The smallest of things could get her sidetracked for fifteen or twenty minutes—sometimes even longer, particularly if it was a rock. She had the wildest of fascinations with them.

"Sorry, Chloe, I found this really cool rock while I was walking around outside." She pulled it out of her pocket and handed it to me. "Isn't it awesome?"

I took the rock and smiled. "Very awesome."

I loved my sister, of that I had no doubt, so whenever she

showed interest in something I would make sure to be as supportive as possible—even when it was with a rock. Admittedly, it was a pretty cool rock. I remember its jet black exterior, and more specifically how smooth and shiny it was. It almost had to have gone through one of those rock polishers at some point.

"Ready to get on the bus?" I asked, handing her back the rock.

"Yup! I'm starving."

"Of course you are," I said, ruffling her hair, which still looked as if she'd just rolled out of bed. "You'd eat until you burst if Mom let you. Come on, let's get going."

We rode the bus home as we always did, nothing of interest to report on the trip. It was a fairly short ride, no more than ten minutes, and then we were back at our dilapidated apartment.

3

As was usually the case, our mom had spent her day doing her absolute best to tidy up the place. The apartment, while in awful shape, was only twenty or so years old. Unfortunately, it was situated in a part of town that catered to low-income families. Low income areas were breeding grounds for high crime and drug use. And those who use drugs don't tend to take care of their things, thus the condition of our apartment.

Despite all of this, we made it work. When entering the apartment, you were first greeted with our kitchen. Its hardwood floors creaked with every other step, and even though our mom cleaned it almost daily, there were some stains that you just couldn't get out. The countertops were eggshell, although they too had stubborn spots that would likely outlive the apartment itself. The cabinet doors were the same eggshell color, and the trim around them matched the floors. Looking back, it was the quintessential apartment kitchen built in the 90s.

As we walked into the door, I saw Mom just getting the table set for an afternoon snack for Char and me. While I didn't necessarily want anything, I couldn't bring myself to say no, especially with how little money our parents had. Char, on the other hand, was pumped. She threw off her bookbag and raced across the creaking hardwood floor to the dining room table. She was about to dig into the snack in front of her, today it was apples with peanut butter, before Mom cleared her throat.

"Miss Charlotte," she said, once again feigning her anger. It was extremely rare, almost unheard of, for her to actually get upset. "What do we do with our stuff when we get home?"

Char sighed, getting up from the table. "We hang it up." She walked back across the kitchen and grabbed her bag, hanging it up on the hook our dad had installed for this very purpose. She ran back over to the table, sitting down again in front of the snack. "Sorry, Mom. I'm just hungry."

She smiled and kissed Char's head. "It's okay, sweetie. Enjoy your snack." She turned to me as I was taking off my shoes and grabbing a glass of water. "How was your day, Chloe?"

One of the many advantages to truly enjoying school was not having to lie when my mom asked about my day. "It was good," I said. "Have you read 1984? By George Orwell?" Again, I knew the answer—at least I was pretty sure. My mom was a wonderful mother, but she wasn't one for the arts.

She shook her head. "No I haven't, sweetie. I've heard of it though." She sat down at the table with Char and me, listening as I rambled on to her about the book. I read every chance I got, and my mom was always willing to listen to me go on and on about them.

Half an hour later my dad, Paul, came home from work. As I said before, he worked for Inforum, a local newspaper. He'd been there for over twenty years, and was one of their senior editors. Unfortunately, the pay still wasn't great, and with only his salary to support the four of us, we were living in near poverty. I can't remember if it was that specific day or not that he shared his big news with us, but it's close enough for the purpose of this story.

"I'm home!" announced my dad as he came into the front door. This was his usual thing, even though it was completely unnecessary in our tiny apartment. We all knew when he got home, but it brought a smile to us all each day. "I've got some exciting news."

My mom, who was washing our dishes from snack time, greeted and kissed him. "Exciting news? What is it?"

He brought my mom over to the dining room table where Char and I were still hanging out. I put my book down and greeted my dad with a genuine smile. He was a great dad, after all.

"So, there's this old lake house on Alpine Lake that Inforum wants to do a full story on. I volunteered to write it, however, I didn't realize what exactly I was in for. Apparently they want to pay for us to live there for a year!"

"A year?" said my mother. "Why on earth for so long? What kind of story is this?"

"They want to capture a first-hand experience of living there. From what I was told, which wasn't much, is that it's been abandoned for a number of years. Can you imagine the history of this place?"

Oh yeah, my dad had a thing for history. Like, an over-the-top thing for it. He always loved when articles in the paper involved anything from the past. He made sure to be the editor to work on them. So, it came as no surprise to anyone when he immediately mentioned the history of such a place.

"I don't know," said my mom. "Where is Alpine Lake? Do you know anything else about this place? What if something awful happened there?"

My dad got up to make a cup of tea—he loved the stuff. "The lake is south west of town. From what I hear, this place is pretty remote, but there's supposed to be a great spot for swimming! It's seated right next to the lake."

"Sounds peaceful," I said. While I wasn't sold on the idea at the time, I couldn't help but be in agreement after seeing the excitement on his face.

Mom finished drying the dishes and joined us at the same, and my dad did the same a moment later with his tea. "It just seems so sudden," she said, scratching the top of her hand. This was something she did whenever she was anxious. "I know you're excited about it dear, and I want to be supportive, I just—"

My dad could tell our mom wasn't on board yet, so he put up a hand to interrupt. "Tell you what, I'll speak to my boss, have him set up a tour of the place for us. Maybe we could spend a weekend there. What do you say, hun?"

"Please, Mom," said Char, who had mostly been looking at her rock collection, but upon mention of it being right by the lake, her ears had perked up. "I bet there's lots of rocks there!"

This got a laugh from both my parents. "Okay, okay," she said. "Let's stay a weekend out there and see how things go.

I'm still not sure how I feel about living somewhere so remote."

"Great! I'll talk with my boss on Monday."

That was the first mention of the lake house on Alpine Lake, at least that I can remember. This was thirteen years ago, after all. Looking back, if either of my parents had a sliver of an idea about what was to come, we wouldn't have stepped foot onto that property. Hindsight is, as they say, 20/20. It certainly rang true in this situation.

After dinner that night, my family all hung out in the living room together. Char was organizing her rock collection, making room for the upcoming weekend trip, while I was reading *1984*. With only thirty pages left, I wasn't about to put it down. Mom and Dad were watching something on tv, probably one of their nightly sitcoms they seemed to enjoy. Nights like these bring a smile to my face as I write about them. Even though we were broke, we were happy together.

4
Therapy Session #249

Dr. Lily Burke had been my therapist for over twelve years. She was a middle-aged woman, single as far as I know, who I can say for certain saved my life on more than one occasion. Her office was always cozy and relaxing. A large stone brick fireplace, flanked by two cherry oak bookshelves, was the focal point of the room. The pleasant crackling from the fire always put me in such a good mood. In front of the fireplace were two oversized chairs, which were some of the most comfortable I'd ever sat in. Dr. Burke always had tea on hand for our sessions, any kind you could imagine. Needless to say, I always looked forward to my time with her.

"So, how are things today, Chloe?" she asked, sipping her tea and turning to a new page in her notebook. "Have you started writing?"

"I have. Unfortunately, I didn't get a whole chapter yet, but I'm happy with what I've written so far. Do you want to read it?"

"Yes, of course," she replied, jotting something down in her notebook. "And please don't be so hard on yourself. Even starting something is an accomplishment."

I handed her what I had written so far, consisting of just over four pages, and waited anxiously for her to say something —anything. Why does time seem to stand still when you're waiting for someone to respond? After a few more minutes, she finally did.

"This is good, Chloe."

I let out a deep sigh of relief. "Oh, thank you," I said, taking a sip of the wonderful tea she had prepared for me. "I honestly wasn't sure where to start. I wasn't even sure if I could start."

She nodded. "That is one of the hardest things for writers to overcome. Once you get going, momentum tends to carry you along. That being said, writing is hard work. Even if you think everything is going well, your characters are developing on their own and almost writing the story for you, writer's block can sneak up on you without a sound."

I wasn't quite sure where Dr. Burke was going with this, however, I learned long ago not to interrupt her tangents—they almost always managed to find their way back to the original point of the conversation. This time was no different.

"What I'm trying to say, Chloe, is that although the idea of writing a novel may seem daunting, you have gotten over the biggest hurdle. You've started."

I smile. "Thanks, Lily." It was a number of years ago when she insisted I call her Lily as opposed to Dr. Burke. It was awkward at first, but now it was like chatting with a friend, which I now realize was the point. "I just hope I can keep going. These times I'm writing about now are easy—they're happy. What about when things aren't so happy?"

She handed me back the printed pages. "I believe in you, Chloe. You're a strong woman, one who saw things at sixteen that someone should never see in their lifetime. You've come a long way since that night in July, and I know you can do this. Trust me when I say that getting it out of your head and onto paper will help."

I knew this was true. Even the little I had written so far seemed to be helping. It was a nice walk down memory lane, and even though there were dark times to come in the walk, it was one I needed to take. "Thanks again for everything," I said, knowing what she would say, but still feeling the need to say it.

"It's my job to help people, Chloe. So, what else would you like to talk about today? We still have thirty-five minutes left of our session."

"This might seem like a silly question, but how detailed should I get with my family's time at…" I had to pause. Just the thought of the word brought back a flood of memories. "At the lake house."

"The more you're able to express, the better this therapy will work. My advice to you is just write what comes naturally when you think about your time there. Don't force anything, just let the memories come."

I nodded in understanding. "Okay, I can do that."

The rest of our session was filled with what could almost be described as small talk. Lily preferred to keep things as light as possible, except on my really bad days, then she would focus hard on the issue that was bothering me. I liked this approach, it was one of the many reasons I kept coming back. The thirty-five minutes passed as quickly as the time always seemed to. Just as we were getting up to walk out, she asked me a final question.

"Have you ever considered going back to visit?"

"What? Going back?"

"I know, I know," she said, putting up a hand as if to say *I can explain*. "It's sort of along the lines of the novel assignment. Sometimes going back to where a traumatic event occurred can actually help. It can give you closure. I just want you to think about it, okay?"

I sighed. "I will." And I planned to think about it, I really did.

CHAPTER TWO

1

The following Saturday after my dad had told us about the lake house, we were packed up and on our way to it. The plan was to check out the place, see how we felt about potentially living there for a year, and hopefully have some fun in the process. It was the 4th of July weekend, too, so we brought some fireworks to shoot off. Char was ecstatic about the idea of a beach full of potentially new additions to her rock collection. I didn't share her level of enthusiasm, but I would have been lying if I said I wasn't intrigued by the mystery surrounding the place.

"Are we there yet?" asked Char, a typical thing of a kid her age to ask. "I'm bored."

I saw my dad look over at my mom and smile. "Not quite, hun, but we're close. Maybe another twenty minutes."

Char sighed, then went back to messing with her rocks. I had noticed almost right away how excited my dad was after confirming our weekend getaway to Alpine Lake. The trip was just over an hour from our house, and he had spent the entirety of it, up until Char had asked about our ETA, going on and on about what he hoped it would be like. My mom was always a good sport about these kinds of things—he was fond of going on long-winded ramblings, especially when it had to do with history of any kind. Just like me with books. Like father, like daughter, I suppose.

"I wonder how many families have lived there?" he asked.

My mom didn't really have an answer. "I guess I don't know, dear. Didn't your boss tell you anything else about it?"

"No, he was quite discrete about the details, actually. I think he wants me to experience it all firsthand without any external biases."

"Hey, Dad?" I asked.

He looked back at me in the rearview mirror, his face showing the pure bliss that had taken over him. "What's up, Chloe? You've been quiet back there. Got your nose in another book?"

I smiled. "You know me." I held up the book I had been reading. If I recall, it was of the Harry Potter books—my third time through them. "Do you think there will be any books at this place? A library?"

I remember him shaking his head and chuckling. "Oh Chloe," he said. "We could be going to the nicest resort on the planet, and you'd wonder about their book selection. Not that that's a bad thing, of course. But unfortunately, I don't know much at all about this place. Literally all my boss told me was where it is, and why he wanted me to stay there. The rest we'll have to find out for ourselves. Five minutes!"

Five minutes later, a scene came into view that would be forever ingrained in my memory. As my dad turned off the main road, we came to the longest driveway I'd ever seen—

even to this day. Looking back, it had to be nearly a mile from start to finish. Lining every few feet of it, on either side, were large eerie pine trees. While I remember thinking of them that way, now that I am writing this, I can't say I know what exactly gave me that first impression.

The latter half of the driveway rose at a subtle incline, and as we reached the top, the house came into view—and what a view it was. The trees separated on either side, creating an opening while wrapping themselves around the rest of the property. This created a particularly comfortable environment, one that could be mistaken as welcoming. The driveway circled around in front of the centerpiece of the property—the house on Alpine Lake.

The front porch, which wrapped around the entire front half of the house, was painted a dark charcoal. While this gave the place quite the first impression, what I remember most vividly was just how pristine the exterior was. If my dad hadn't told us it had been sitting abandoned for nearly a hundred years, I would have never known it. The paint, the wood that the porch was constructed of, even the grass in the front yard—it was all perfect.

Too perfect.

Dad stopped the car right in front of the house, a for sale sign was stuck into the ground next to the driveway. I doubted anyone even knew this place existed, let alone that it was for sale.

The house itself was painted a hunter green, and like the porch it was in immaculate condition. My eye had been drawn to the front door, which was painted the most brilliant white I'd ever seen. I remember the sight of that door sending the most unsettling chill through my entire body. Perhaps I should have taken that as a bad omen, but at the time I was young and just didn't think of it that way.

"Would you look at this place," my dad said, practically jumping out of the car. "It's beautiful!" He stood next to the for sale sign and took in the view of the house.

"It looks so new," said my mom. "Are you sure we should be staying here? What if it's not abandoned?" My feelings on the place were somewhere in between my mom and dad. While I was excited to see what the inside was like—and if there was a library—I couldn't help but feel something wasn't quite right about the situation.

My dad was already unlocking the front door, but turned around to face us. "Oh it'll be fine, dear," he said, opening the door and letting it swing to the right. As it did, we were greeted with our first view of the house's interior. "Whoa," said my dad, taking the first step inside. "You guys better come take a look at this!"

While my mom helped Char lug her rock collection out of the car, I carried my weekend bag and current HP book up towards the porch. As I put my foot on the first step, I felt the board beneath me bend as it took on my weight. A second later it stopped, and I continued on until I was standing inside the entryway next to my dad. It took no time at all to understand what my dad was talking about.

"It's so clean," I said, taking a few more steps inside. Clean was an understatement. The place looked immaculate, which did nothing to help alleviate the weird vibes I was getting from the house. Immediately to the left of the entryway was a staircase leading to the second floor. I ran my hand along one of the banisters, and when I turned over my hand I didn't see a speck of dust. "I thought this place was abandoned, Dad."

He nodded. "It is. From what little I found about this

place, the last family to live here was nearly one hundred years ago. But there's no way a place could stay this clean for even close to that long."

My mom and sister finally made it up to the porch, and joined us inside. Their reactions were nearly identical to ours. Opposite the staircase was a bedroom, although it appeared to be set up as an office. As I walked farther in and made it to the kitchen, I continued to be surprised at how little dust or dirt there was. Mom kept our apartment clean, nobody would question that, but this place was on another level.

"I don't know if I like this," said my mom, setting the bags of food she had brought with onto the kitchen island. "Are you really sure nobody lives here?"

Dad shrugged. "I know as much as you, dear. Maybe my boss hired someone to clean up prior to our arrival." He ran his hands through his hair as he thought over this possibility. "Yeah, that must be it. I'll have to give him a call and say thanks." My mom didn't look entirely convinced, but she let it go. "Come on, let's go check out the rest of the place."

2
Therapy Session #250

Normally I meet with Lily every two weeks, however, since I started to write about the first time we visited the house on Alpine Lake, I thought it was best to see her asap. As always, she brought me back to her office with a smile and two cups of tea sitting between our chairs. I sat down and grabbed my cup while she picked up her notepad.

"So, you've reached the lake house," said Lily. "How are you feeling about it?"

I take a sip of the tea, which is as delicious as ever, and

think about her question. "I'm not sure, honestly. It's not bad yet, but I can feel the darkness creeping up in my mind. Right now it's okay because I know we're just visiting for the weekend. I know nothing—" I stop mid sentence as an image storms into the forefront of my mind. I shudder, trying to push it away. "I know nothing too scary happens during that weekend, but…"

"You know it's going to get worse," she said, finishing my thought. I nodded. "That's to be expected, Chloe. It's going to be scary revisiting these memories, but you can do this. Just remember that you're safe now, and all you're doing is writing down the memories for some sense of closure."

"Right," I said, although not very convincingly, which of course she picks up on.

"I know, easier said than done, right?"

I nodded again. "I'm terrified, to be honest. What if I can't finish the story?"

Lily smiled. "Then you don't finish it. At least you can say you tried."

We sat in silence for a while after that. The fireplace was lit, and the crackling of the wood was helping to calm me down—it always did. I took another sip of tea, chamomile with honey today, and let out a heavy sigh.

"You're right, of course," I said. "Thank you, Lily. I am actually feeling quite a bit better about continuing on with the story."

"Great! I'm looking forward to reading more of it. Is there anything else you'd like to discuss today?"

"I don't think so. I'm ready to get back to writing my story. I plan to have two more chapters ready for you by the time we meet again."

Lily jotted a note down, I assumed something about

holding me to my two chapters, and smiled again. "Sounds great, Chloe. I'll walk you out then."

I always feel better after our sessions. Even when they're only ten or fifteen minutes long, just being able to openly talk about anything that's on my mind is beneficial. It helps knowing that Lily is a safe person to talk to, and I know she'll never judge me—and yes, I know that's her job—but it really does help me.

3

While the house's kitchen was impressive in its own right, where the place really shined was the backyard. The sliding glass doors opened up to a large deck surrounding the entire back half of the house. The building was up on a hill, so the back of the property slanted down towards the lake. From the deck you got a bird's eye view of the tree line, and just beyond that, a scenic view of Alpine Lake. The house was close enough to hear the water lapping against the beach on a windy day. I remember thinking the view belonged in a romance novel or something—not that I was really into that kind of story, but it's what first came to my mind.

Below the deck was the backyard, which admittedly wasn't anything to write home about (there's a joke or pun in there somewhere), but as your eye went to the tree line, a small break in it revealed a dirt path leading down to the lake. I had just sat down on one of the chairs placed against the outer wall of the house when my mom came out through the door.

"Wow," said my mom as she joined me on the deck. "This view alone might change my mind on this place." She walked over to the edge of the railing facing the lake, which looked to be impossibly close to the house. "Want to go check out the beach, Chloe?"

I hesitated a moment, looking back at my unfinished book, but decided I should try to enjoy other things as well. I nodded. "Sure, might as well while we're here, right?"

I put my book down and went with my mom down the stairs towards the overgrown grass of the yard. We quickly reached the break in the tree line and walked down the short dirt path. At the end of it we got our first look at the beach, which was littered with rocks and various debris from the lake. My mom seemed disappointed, and while I was disinterested in the situation as a whole, I knew Char would be thrilled at the sight of all these new additions to her rock collection.

My mom read my thoughts. "Well, Char is going to love this," she said. "I guess I shouldn't have expected a sandy resort given the surroundings, huh?"

"Speaking of, what are Dad and Char up to?" I asked, picking up one of the nearby rocks and looking it over.

"Char insisted on unpacking all her rocks, even though we're only here for the weekend, so he's helping with that. Maybe we should go back and save him." She laughed. "What do you think? Not much else to see down here."

I agreed. It really was a sad looking "beach", but as my mom had said, this isn't some private resort we were staying at. Private, yes, but resort? No, not even close. "Yeah, let's head back. Maybe I can take Char back down here and we can get her some new rocks."

"That sounds great," she said. "I'm sure she'd love that." She put her arm around me as we made our way back up the dirt path and into the house.

4

Twenty minutes later, Char and I went down to the lake in search of rocks. As expected, she was over the moon at the sight of the "beach". While they'd never admit it, I think my parents were tired of all the rocks, so I did my best to be the supportive big sister. I stood next to a chunk of driftwood as Char went from rock to rock, inspecting every minute detail of them.

"Oh, oh," she said, apparently finding a super cool one. "Chloe, look at this one!" She held up a rock that had been smoothed by the years of water washing over it. "It's so smooth."

I can't help but smile. "It sure is, Char. What do you think of this place so far?"

"It's great! I don't think I'll have enough room for all these new rocks though."

This makes me laugh. "You can't take them all, silly. You'll just have to choose a few you really like. What about the house? Would you like us to stay here for a whole year?" Part of me thinks she missed my questions and was focused on not being able to take all of these rocks back home.

A moment later, though, she managed to respond. "I don't think so. It's creepy." That was all she said at first, which was less than comforting to me. I felt that unsettling feeling return from earlier, when I first saw the house's front door.

"Why do you say that?"

Before I could get an answer, I heard our mom calling us back up for lunch. Char quickly grabbed two final rocks and we walked hand-in-hand back up the dirt path to the back-yard. As I looked up towards the second story of the cabin, a sudden movement caught my attention. It was in one of the upstairs windows—the dark drapes covering it had moved, as if someone had been looking out at them. That unsettling feeling came again—something just didn't feel right about this place.

5

The rest of our weekend at Alpine Lake was mostly uneventful, but surprisingly pleasant. Upon exploring the house further, I found four bedrooms, three bathrooms, and to my utmost delight—a library. While it wasn't on a grand scale such as ones requiring a ladder on wheels to access the upper shelves, it was nevertheless a room filled with books just waiting to be read. At first glance, there must have been at least a dozen bookshelves, each with six or eight shelves filled to the brim with books.

The library hadn't been easy to find, however. I was in one of the two upstairs bedrooms, which was the larger of them, and I heard a subtle noise coming from the far wall opposite the bed. I remember it being a scratching sound—like something trying to get in—or out. As I moved closer to the sound, I noticed a small outline in the wall. It was barely noticeable, not something you'd pay any mind to unless you were drawn to it like I had been that morning. As I pressed my ear against the wall, the scratching stopped. The wall suddenly began to rotate. A moment later I was standing inside a hidden room—a hidden library.

Unfortunately, I made the discovery of this library not long before we had to leave, so I was only able to scan a few shelves of books. A few caught my eye, and after seeing all that were left, my opinion of the house began to change. The truth was, I wanted to come back and stay for the year—or longer. I had managed to completely put the scratching sound out of my mind. I had found a treasure trove of books. When my mom came to get me that Sunday afternoon, I had just pulled one of the books from the shelf—it was *Dracula* by Bram Stoker. The book was old, older than any I had seen before, and I asked if I could take it home.

She smiled and nodded. "Of course. Come on, we need to get going though. Your father is anxious to get back and discuss the details of our year-long stay here."

My ears perked up at this. "So, we're going to come back?" Before finding the library, I couldn't be sure if I would have wanted to ever see this place again. Funny how a room full of books can make even a creepy house by the lake seem lovely.

"Well, it isn't for sure yet, but your father seems dead set on the deal. We'll discuss it as a family once we get back home. Ready to go?"

"Yeah, I packed up this morning, I just need to take my things downstairs."

"Sounds good. I'll help. Char and her rocks are already loaded in the car."

My mom and I take my things downstairs and out the door to the car. Looking out the window, I see my dad carrying the rest of their bags out. He sets them on the porch and locks the door to the cabin, bringing our weekend stay to an end. He piles the bags in the trunk and gets into the driver's seat.

"Everyone buckled?" he asked, starting the car and putting it into gear. As our car pulls around the circled driveway, I take one last glance at the house. Another brief movement catches my eye, this time from the bedroom window of the room I'd found the library in. It was only the drapes swaying again, but I could have sworn I saw a hand—if only for a second.

CHAPTER THREE

1

As I'm preparing to write this next section, I wanted to take just a few moments to get my thoughts in order, and as the reader you get to come along for that—lucky you. The days leading up to us moving out to the lake house are some of my fondest memories of that year. For a while I wondered about this, but now it makes sense. Those days were full of excitement—I wanted to read every dust covered book in that secret library I had found. Spoilers, I didn't get to read them all, but the ones I did read were some of the best books I've laid my eyes on, even to this day.

I know what you're thinking—what kind of sixteen-year-old is obsessed with reading old books she found in a creepy library? Times were different back then. While the internet was starting to get big in 2010, it was a far cry from what it is today. But even so, my family was dirt poor. While we never went hungry or anything like that, there just wasn't money for things outside of the bare necessities of life—food, water, shelter, etc. Books were cheap, free if I got them from the local library, so I got myself lost in stories whenever I could.

Anyway, with all that said, I'll let you get back to this story. Hopefully you're finding yourself lost in this one as well.

2
July 5th, 2010

The Monday after we got back from the lake house was one of the longest days of my life. I sat in school watching the clock tick by, anxiously waiting to go home and find out what my dad heard from his boss. He was going to discuss the details of us living there for a year, and I wanted more than anything to go back there—just for that library. I was obsessed with it. After what felt like an eternity, the final bell of the day rang, and I was free.

Ten minutes later, I was standing in our usual meeting spot, waiting for Char. She came along a few seconds later, no rocks in tow to my surprise.

"No rocks today, huh?"

She shook her head. "I'm trying to save room for those rocks at the beach. Do you know if we're going back?"

I smiled at her and ruffled her hair, which was a tangled blonde mess, despite my mom's best attempt at making it presentable that morning. "I sure hope so, Char. I've been going crazy all day wondering the same thing. Come on, let's get on the bus so we can get home. I think Dad said he'd be home early today."

The ride home was thankfully short, and as we stepped off the bus, both of us ran inside and up the stairs to our third floor apartment. As we burst through the door, Mom and Dad were sitting at the table with coffee in hand. I knew what they were talking about, of course, and based on my dad's expression, the news was good.

"Hi!" I said, throwing off my shoes and joining them at the table. "What'd you find out, Dad?"

He chuckled at my obvious excitement. "I should have known you'd be wanting an update. Mom told me about that library you found. Well, you're in luck kiddo, we've got the place for a year!"

"Yes!" I shouted. "When can we go?" Char had finally got her jacket and school bag put away, and she joined us at the table.

"More rocks!" was her contribution.

"Take it easy you two," said our mom. "We have to pack up some things before we can go as well as make a few arrangements for while we're gone. A year is a long time, after all."

Truth be told, I hadn't really thought about that part. Not surprising, all things considered. Nevertheless, she was right. At that moment, something else dawned on me that was also lost in the excitement of that library.

"Hey, Mom," I said. "What about school? Will we still be going while at the lake house?"

My parents exchanged a glance, which may have gone unnoticed by Char, but I knew it to mean this was a topic they had recently discussed and were still not entirely decided on yet. That was something I came to admire about them—no matter how hard the conversation was, if it needed to happen, they made sure it did in the most civil manner possible.

Finally, my mom answered. "Your dad and I were discussing that just before you two got home, actually. We were thinking of doing a year of homeschooling while we're there. It would be an adjustment, but I think it would work out well for everyone. During the day while your dad is working on his writing, the three of us can do the school thing. What do you think, Chloe?"

Something else I loved about my parents—they always wanted to know what we thought of an idea. If we didn't agree with something, or didn't know for sure if it was what we wanted to do, they would hold off on moving forward with it until we could discuss it further as a family. There were exceptions, of course. Sometimes life just happened a certain way and you had to deal with it.

"Could my friends still visit?" I asked, thinking about the few but close friends I had made over the years in school. "If so, then I'd be okay giving it a try."

My mom smiled. "Of course, dear. That was one of the sticking points we had during our discussion." She put her arm around me and pulled me in for a hug. "I know you keep your circle small and close, so the last thing I'd want to do is disrupt that at all. Homeschooling next year then?" Both me and my sister nodded, the latter of whom had mostly been thinking of all the new rocks she'd have to collect on the 'beach'.

"Great!" said my dad. "Now, how about we get some dinner going and then we can discuss the plan for packing and moving."

The rest of the evening went as I expected it would. We ate dinner, spaghetti with garlic bread, and then the four of us sat in the living room making lists of things we wanted to bring to the lake house. We were only planning to take any personal items to the lake house as my dad had said his boss was providing us with funding to purchase any food and supplies we needed during our year-long stay. This had been a huge relief to both of my parents for obvious reasons.

As evening turned to night, I said goodnight to my parents and sister, and then sat in my bed to do some reading. For one of the first times in I think ever, I found myself unable to concentrate on the words on the pages. Funny enough, it was a library full of books that was distracting me from the current book I was trying to read. After a few more minutes, I managed to push the thought away enough to read a few chapters before turning in for the night. I couldn't wait to spend a year at the lake house—with all those dusty old books.

3
Therapy Session #251

I went a month this time without scheduling an appointment with Lily. For once in, I honestly can't remember how long, things have been going okay for me. As mentioned earlier, the weeks leading up to us moving to the lake house (as well as the early days there) are some of the happiest memories I have. While I am not at all looking forward to what I know is to come, assuming I manage to write the whole story, at least part of me feels like I may finally be making steps toward moving forward with my life.

As always, Lily greeted me with a warm smile, holding a cup of tea. She motioned for me to come back to her office where I know there will be a cup of tea for myself. Routine is comforting, more so after everything that happened at the lake house. I like knowing what is going to happen, and Lily is well aware of this—probably part of the reason she does what she does.

"Good morning, Chloe," she says, her genuine smile never leaving her face. "It's been a little while. How are you doing?"

I take a sip of tea, Earl Grey today (my favorite), and respond as honestly as I can. "I think I'm doing okay today."

Lily jots something down in her notebook. "I'm glad to hear that. So, you think the writing is helping?"

I nod. "Oh yes, most definitely… at least, for now."

"Right, you did mention that you're currently at a particularly happy part in the story. Did you bring me your most recent work?"

I had brought it, of course, but I was hesitant to let her read it. This would be the first portion of the writing that she read which would include these snippets of our sessions. While I didn't feel it was entirely necessary, I felt bad about not asking for her permission before including them—not that I couldn't just as easily remove them after the fact if she said no.

"Here you go," I say, reluctantly handing her the printed out pages I had brought in. "Sorry if it isn't as good as my first few pages." I take another sip of tea and find myself zoning out into the fireplace, which is as comforting as it usually is. I wonder what it is about a warm, crackling fire that makes us so at ease?

"This is good," she says, flipping back through the pages. "Are you planning to include more of our sessions?"

"I—uh, I would like to. If that's okay with you of course."

She smiles and hands me back the papers. "Of course. I think it adds a nice touch to the overall story… makes it a little more real. I'm glad you enjoy the tea and fireplace so much, by the way." That smile again—it really is quite nice. I feel my face flush as I try to hide behind my cup.

"Thank you, Lily. That means a lot."

She jots something else down in her notebook. "So, there's something I'd like to try, Chloe. It doesn't have to be this session, in fact it might be best if I tell you what it is and then you can mentally prepare for it for the next time. Sound fair?" I nod. "I'd like to go back to the night the police found you on the porch. We've tiptoed around this for years, but I really think it's time to talk about it. I was thinking we'd schedule an extended session so you can take as long as you need. And before you say yes or no, just give it some thought okay? You can get back to me before next month."

"Sure," I say, not really giving myself time to think about it. "I can do that."

One more warm smile appears on her face as she sets her notebook down and stands up. "Well, I think we can call this session done then. Shall I walk you out?" I agree and follow her out of the office. My mind immediately starts racing, and I am already regretting what I agreed to.

4
July 10th, 2010

It's nine in the morning, and the four of us were already packed up and on the road—off to the lake house for a year. We had spent the week packing our things, making arrangements for a friend of my dad to look after the apartment, and also informing the school that me and my sister would be taking the next year off. The principal wasn't thrilled about the idea, but it was ultimately my parents' choice to do what they wanted with our education.

I spent the drive reading the last few chapters of *Dracula*, the book I took from the hidden library at the lake house. Normally I would have been done with a book of this length in a day or two at most, but with everything going on regarding the move, my mind had been distracted. Char, who has never been a morning person, was passed out in the seat next to me—currently snoring against my shoulder.

"So," said my dad, looking back at me in the rearview mirror. "I know what you're most looking forward to, Chloe, but is there anything aside from this hidden library that you're excited about?"

I thought about it for a moment, not wanting to say my first thought—which was 'no'. If it was Mom asking, and Dad wasn't around, I probably would have been honest, but I know how important this is to him.

"I'm looking forward to spending more time with you two. Usually school takes up most of the days, but now we'll be home together way more." I saw him smile, and it was certainly genuine. He loved his family.

"That will be nice," he said, placing a hand on my mom's leg. "I'm looking forward to exploring this place more, finding out everything I can for my story. What about you, dear?"

I didn't wait to hear more. I put my headphones in and got back to reading. Only one chapter left, and if memory serves me, we had about thirty minutes left until we got to the house—plenty of time to finish.

5

Thirty minutes later I'm once again greeted by the sight of that impossibly long driveway flanked by thick pine trees. As we approach the house and my dad parks in front of it, a chill goes through my entire body. I don't like the feeling, but fortunately once it's gone I feel okay again. Char had just woken up from her nap a few minutes ago, and was groggily looking out the window.

"We're here!" she shouted, unbuckling herself and opening the door. "I can't wait to find new rocks!"

The rest of us got out of the car and opened the trunk to start carrying in our things. I take my two bags—one with clothes and toiletries, and one with a few of my favorite books that I simply couldn't part with. After my dad opened the front door, I went up to the room I had chosen last time to be mine and started to unpack. Like the kitchen, the bedrooms had all been recently cleaned with no sign of even the smallest speck of dust.

Immediately to the left of the door was a large chest of drawers. There was plenty of room for the minimal amount of clothing I had brought. The full sized bed was against the adjacent wall, and across from that was the wall housing the hidden library. The walls of the room were painted a warm tan color, and the floor was covered with a lush beige carpet. I thought it was a nice room, although at sixteen I wasn't particularly picky. I had my own room for a whole year—I would have been happy with pretty much anything.

Across the hall I heard Char unpacking all of her rocks, the sound of them clanking against one another was unmistakable to me at this point. Before going to see her progress, I wanted to get myself a new book to start. I walked over to the wall with the hidden door and looked for the opening. Since I had recently opened it, finding the spot where it separated was much easier than the first. I pulled open the door and was met with the distinctive smell of old books—there really isn't a good way to describe it, but if you've experienced it, you know.

The hidden library room was about seven feet across, and it was maybe a bit deeper. The middle walkway was flanked on either side by three rows of bookshelves stretching to the ceiling. A single bulb hung in the center of the room, and even when lit barely illuminated the room. While it certainly added to the creepy factor, it made it difficult to read the book titles. I put *Dracula* back where I found it and began looking for my next book.

I started with the bookshelf that *Dracula* had been on, which contained a few classics I was actually familiar with: *Catcher in the Rye, 1984, To Kill a Mockingbird, etc.* I'd read all of these before, but seeing these particular copies was quite a thrill. All of them appeared to be first editions—the original printings of the book. *Someone started this collection a long time ago,* I thought.

"Over here," whispered a voice. I jumped at the sound, dropping a dusty book I was holding onto the floor. I felt my heart rate start to increase.

"Wh—who's there?" I asked, not particularly wanting a response. After not getting one, I moved farther into the room, standing in front of the second bookshelf on the right. I had thought this was where the whisper came from.

"Closer," whispered the voice again, sending chills down my spine. "Almost there."

"Chloe? Char?" said a familiar voice. "Can you two come down here please?" It was my mom. I decided I would save picking another book for later and dashed out of the library, down the stairs, and met my mom in the kitchen. Char came down a moment later, a rock in tow of course—one of her favorites.

I was out of breath by the time I got downstairs, and I could feel my heart was still racing—both from the stairs and whatever was going on up in the library. While I hoped it wouldn't show, my mom's face told me it did. I didn't like to worry my parents, even though I'm certain it was something my mom always did.

"Are you okay?" she asked. "You look like you've seen a ghost."

"Yeah, I'm fine. I just finished putting my stuff away. What's up?" I could tell she didn't like the answer, but she let it go.

"How are you liking the room? Do you need help with anything?"

My mind went back to that voice—I know I heard it, but now wasn't the time to tell her. She'd probably just think I was adjusting to the new place or something.

"The room is nice," I said. "Nice to have my own space."

She smiled. "I figured you'd enjoy that. What do you guys want for lunch? It looks like Dad's work really went all out to supply us with food. We've got a little bit of everything."

"I want chicken nuggets," said Char. "And french fries!"

My mom smiled and shook her head. "Of all the things. Chloe, is that okay with you? I don't mind making something separate if not."

Again, something I'd always love and remember about my parents—their unrelenting love and compassion for both my sister and me. Sometimes it was something small, like offering to make the two of us separate lunches, but it was a constant in my life, and I came to appreciate it more and more.

"Chicken nuggets and fries sounds fine," I said, grabbing a bottle of water from the fridge. "Where's Dad?"

"Oh, he realized we forgot a few things back at the apartment, so he's headed back to get them."

The rest of the afternoon was pretty chill. The three of us ate chicken nuggets and fries, and afterward we took a walk down to the beach. It was a beautiful summer day—perfect for being outside. My mom and I sat in two of the lawn chairs we had taken from our apartment while Char did her rock scavenging. Because of the incident in the library earlier, I didn't have a book with me, so I just made small talk with Mom. We discussed our plans for our time here, including the whole homeschooling thing and whether or not she felt up to it. I felt optimistic about it, and assuming the voice I had heard earlier was just in my head, I was looking forward to the year at the lake house.

6

Over the next week, we finished unpacking all our things, and really started to make the lake house our own. Dad got his office set up, which is where he'd be spending the majority of his time during the day. Mom was busy as ever hanging up things on the walls, putting pictures of us in frames throughout the house, and always making sure we had plenty to eat. She was ecstatic about the amount of food available—truth is, we all were. I can't remember a time when we had this many options of what to eat. It was, in a way, a bit overwhelming.

I decided to hold off on going back to the hidden library—at least for a bit. I had brought a number of books with me from home, a few of which I hadn't read yet, so there was no hurry to go back in there. I was currently reading *The Shining* by Stephen King, which seemed fitting. The characters in this book are staying at a remote hotel while we're at a remote lake house. I wasn't too far into it yet, but I was enjoying it.

As I was laying in bed reading one afternoon, Char came into my room with a book of her own. This caught me off guard as she wasn't one for reading, but I was certainly going to support her sudden interest, if that's what it was.

"Hey, whatcha got there?" I asked, saving my place in *The Shining*.

She handed me the book, which I immediately noticed was very old. "I just found it in my room. I don't like it… it's scary."

I took the book from her and looked at the cover, which had a creepy old house on it. It wasn't the same as the one we were staying in, but there were definitely similarities. And for someone her age, it would be scary. The house on the cover was rundown and looked to be covered in a black slimy substance. I had to admit the sight of it made me shudder. The title was *A Night at the Lake*. Oddly, there was no mention of the author anywhere.

"Where'd you get this?" I asked, flipping through the first few pages, trying to find out more about the book.

"It was in my room," she said again. "After we came back upstairs from dinner I found it on my bed. But it wasn't there before, I swear."

The book suddenly began to shake in my hands. I dropped it in a panic, and one of the pages cut my finger.

"Ouch," I cried as blood began to appear on my index finger. Char had taken a step away from the book.

"Chloe, you're bleeding!"

"It's okay, just a papercut. I'm fine."

"What's going on with that book? I saw it move on its own. I know I did."

I nodded. "Yeah, I saw it too… and felt it." I went to the bathroom to wash the cut and then came back to my room. Char was still keeping her distance from the book, and she was thankful that I was back.

"Should we tell Mom and Dad?" she asked.

I shook my head. "I don't think they'd believe us. Not without more proof anyway." I reached down and slowly grabbed the book off the floor. It felt oddly heavier than a few moments ago, but nothing happened when I picked it up. I handed it to Char, but she backed away and shook her head.

"No, you keep it, Chloe," she said. "I'll stick with my rocks." She turned and ran back to her room, leaving me alone with the creepy book.

I sat on my bed for a while, thinking of what I should do with the book. Despite still feeling hesitant about the library, I decided to face my fears. I got up and found the crease in the wall, pried open the hidden door, turned on the single light, and stepped into the musty, dimly lit room. I quickly found the spot where the book belonged and put it back on the shelf. A moment later, I heard the voice again.

"This way," it said in a whisper. The voice was to my left, from the opposite bookshelf that I was in front of.

"Wh—who are you? What are you?" At first there was no response. "Hello?"

"I need to talk to you," the voice finally replied.

Another chill ran down my spine as I realized that this voice wasn't in my mind—it was actually happening. *What should I do? Talk to Mom and Dad? They'd never believe me.* I decided now wasn't the time, and grabbed a random book from the shelf in front of me and ran out of the library, closing the door behind me. As I was closing it, I heard the faint sound of laughter coming from the left side of the room, followed by the scratching sound again.

CHAPTER FOUR

1
Therapy Session #252

So, reader, do you remember what I was supposed to think about since my last session? I sure do. Lily wants to go back to the morning after it happened—after my entire life was turned upside down at age sixteen. I'd be lying if I said I don't remember that morning, however, I don't remember anything about what happened just before the police showed up. This is one of the blank spots in my memories I mentioned earlier. For a solid twelve hours, I don't remember anything. Perhaps the more I write, the details will come back. Part of me hopes they do, and part of me hopes they don't. I guess if you're reading this, it's safe to assume at least enough details came back to make a book.

I arrived at Lily's office at 8:45. It was bitterly cold, and I had thought on more than one occasion of cancelling this appointment. Needless to say, it wasn't only because of the cold, but it certainly was a factor. A few minutes after I arrived in the waiting room, Lily greeted me with her warm smile and walked me back to her office. As always, a piping hot cup of tea awaited me—it was even better on a cold morning such as today.

"Good morning, Chloe," she said, picking up her notebook. "How are you today? Quite chilly outside."

"Good morning. I'm doing okay. And yes, it's dreadful, but this tea is lovely."

She nodded and smiled. "So, do you remember what we discussed last month? I hadn't heard anything from you, so I wasn't sure I'd see you today."

I remember being somewhat taken aback by her jumping right into things, but I should have expected this, right? I've only been putting off this very conversation for twelve years.

I sighed before responding. "I remember very well, Lily. We can try to discuss that morning, but I can't promise I'll be able to make it through our entire session."

"Of course," she said. "I wouldn't expect you to promise something like that. Just take it from the beginning, or wherever you'd prefer to start, and tell me anything you can. Anything at all will help."

As it turns out, I remembered far more of that morning than I thought. Once I started rambling, I went on for almost an entire hour. Quite a few tears were shed throughout, but to my surprise I managed it pretty well. Lily finally interrupted me, saying that we should wrap things up. I saw the time and couldn't believe it.

"Jesus, I'm so sorry. I didn't—"

She put up her hand. "Chloe, don't apologize. This has been great. I'm quite impressed with your recall. One thing I wanted to ask. This journal… you mentioned it several times. Do you still have it?"

I shook my head. "Oh no, I haven't seen that thing in years. Honestly, the police might have kept it after their investigation."

Lily jotted something down in her notebook and got up to walk me out. I'm not sure why I felt the need to lie to her during that session. It was the first and only time I did, but something about that day made me worried, and I just instinctively said I no longer had the journal.

"Have a good rest of your day, Chloe," Lily said, closing the door to her office as I walked out through the waiting room.

"You too, Lily."

It only seems fitting to write about the day I discovered the journal, right? I remember that day very well—sometimes a little too well. It wasn't long after hearing the voice inside the hidden library, a week later at most. There are definitely times I wish I never found it—oh how differently things would have been had I not.

2
July 16th, 2010

The following Friday was a beautiful day at the lake house. The sun was bright in the clear sky, birds were chirping, and everyone was having a good time. With my parent's permission, I decided my room could use a little color, so I spent the day painting. I had the window in my room open since it was so nice out, which was bringing in a light breeze that felt good while I worked. Blue was my favorite color, so I was doing several shades of it on all four walls. I know my mom questioned the choice, but ultimately told me it was my room to paint how I wished.

It was when I moved the bed out from the way to paint behind it that I saw a chunk of the floor that didn't look quite right. It was subtle—just a small cutout of the floor right next to where it met the wall. You would only notice it if you were down on your hands and knees like I was, painting. I put down my brush and as I put my hand on the spot of the floor, I immediately felt it was loose. With minimal effort I pulled out the loose chunk of wood, revealing a tiny hidden compartment.

Before doing anything else, I got up and closed my door, making sure to lock it as well. My parents weren't ones for snooping, quite the opposite, but I had no idea what I might find in this hidden compartment. I sat back down on the floor next to it and reached my hand inside. What came out was an old leather bound journal. It was covered in a heavy layer of dust, and its age was made obvious by the wearing of the leather, particularly on the strap that held it shut. On the front of the journal were the initials *J.A.* I gasped at the sight of it. *Imagine what's written in here,* I thought to myself. *I have to read it!*

I sat on my bed with the journal and looked at it from every angle. I was mesmerized by the thing. I had no idea how old it was, but based on its appearance, it had to be from when this place was first built. I carefully unbuckled the clasp and opened it up to the first page. It apparently belonged to a girl named Jessica Ambrose, which lined up with the initials on the front. Turning the page I was met with the very first entry.

June 23rd, 1912

My name is Jessica Ambrose. I'm fourteen years old, and this is my first journal entry. My family recently moved into a new house on Alpine Lake. It's only been a week, but so far we love it here. My dad moved us here for his work... he's a lumberjack, and he works in the forest nearby. While he works, my mom stays home with me and my younger brother, Jack, who's nine. If you're reading this, and it better not be you, Jack, then I hope you enjoy reading about my day-to-day life, but I fear it might be a bit boring most days. I think that's it for now. Mom is calling for me to help her bake cookies. Bye for now.

I remember being awestruck at the words I had just read. What are the odds I would just happen to move the bed and find some girl's journal from nearly 100 years ago? I reread the first entry and noticed how similar Jessica's family was to my own. A mom who stayed at home, and a younger sibling, although she had a brother, and a dad who had moved them to this house for work. I knew I had to read more, but I wanted to avoid having to hurriedly hide it if anyone were to come in while I was reading, so I tucked the journal back into the hidden compartment and got back to painting. Those words from Jessica would remain at the forefront of my mind the entire afternoon. I couldn't wait to find out more about her.

3

During dinner that evening, as expected, my mind was preoccupied with reading more of Jessica's journal. Part of me felt weird about reading it at all, kind of like I was invading her privacy, but at the same time I felt I needed to know what she experienced. Despite this, I managed to make it through the meal without drawing any suspicion from my parents. My mom had made tater tot hot dish, which was both my sister and I's favorite, so that helped with providing some distraction from the journal.

After we finished eating, I excused myself and went back to my room. Thankfully, it was nothing new to my parents that I wanted to go back to reading, so again they weren't at all suspicious. I knew at some point they'd likely catch me reading the journal, but I figured I would deal with that when or if it came to be. In the meantime, I planned to find out as much as I could about Jessica and her family. I quietly shut my door, pulled out the journal from underneath the floor, and opened it to the next entry. It was dated the next day.

June 24th, 1912

Making cookies with Mom was great! I love baking with her, and she always makes the best treats. Dad says his first week of work was good. My little brother and I are being home-schooled, and Mom is very strict about making sure we finish all our studies. I'm a little sad about having to leave my friends from my old school, but Mom says we'll only be living here for 6 months, so I'm trying to make the best of it.

I've been reading a lot lately. Oh! Also, my dad said he would be building me my very own library! I'm so excited! He's going to build it into one of the walls of my room, and he's even going to put in a hidden door. How wonderful is that? I can't wait to fill it with all of my favorite books. I think that's all I have to report right now. I think I'll do some reading and then turn in for the night. Good night, journal.

"So that's how the library came to be," I said to myself, closing the journal. "What a cool thing to do for your daughter."

I spent the rest of the evening reading a few chapters in my current book. It was close to midnight when I finally pulled myself away and tried to get some sleep. Just as I was drifting off, I heard the scratching sound again—louder than before. I was instantly wide awake, my heart racing in my chest. *What is that?* I needed to find out.

I got up and moved toward the library door, and just as I reached it, the scratching stopped. *I don't like this.* I took a deep breath and leaned my ear against the door, listening.

Nothing.

I went to move away, but then I heard it—the voice. "Chloooooeee."

I gasped, stumbling back from the door. *What the—* "Wh—who's there?"

Nothing, again.

You're losing it, Chloe. Go back to bed.

After a few more seconds, I laid back down and tried to forget about it. Not long after my head hit the pillow, the scratching started again, but this time it sounded much closer to me. I couldn't bring myself to get up, instead I covered my head with the blanket and held my breath. The scratching stopped, and I heard what sounded like the flipping of pages in a book. *I don't know what to do.*

A moment later, the room went quiet enough to hear a pin drop, which was somehow worse. I somehow managed to fall asleep after what felt like ages, and thankfully the rest of the night was normal.

When I woke the next morning, I thought for sure last night had all been a bad dream, but when I got out of bed I noticed there were bits of torn up paper on the floor. The scraps started right next to my bed, and as I followed them they went across my room and over to the library door. Panic fell over me again—it was real, all of it. *I need to tell Mom and Dad.* I quickly but quietly got out of bed and went downstairs.

4

I found my parents sitting at the dining room table, having coffee and donuts. Apparently my dad had gone out earlier to get some for us for breakfast. My mom said good morning and asked how I slept, but once again she saw something was wrong. I didn't plan to deny it this time.

"What's wrong, Chloe? You look like you've seen a ghost."

Maybe so, I thought. "I—I don't know," I said. "But I think someone was in my room."

"What do you mean?" asked my dad. "How could someone be in your room and you not know about it?"

I shook my head. "Could you guys just come upstairs for a minute? Please?"

Seeing the dread in my face, my parents both agreed and followed me upstairs to my room. Once inside I immediately noticed the paper was gone entirely—no trace of them at all. *What the heck? I know I wasn't dreaming.*

"I swear there were scraps of paper all across my room!" I walked towards the hidden door, inspecting the floor but finding nothing. "I don't understand, they were right here!"

"Maybe it was just a nightmare," said my mom. "You do read a lot of scary novels."

"No, I was definitely awake," I said, feeling my confidence wane by the second.

My dad canvassed the entire room and then turned back to us. "Well, I don't see anyone, nor do I see any signs of them being here."

A few moments of silence filled the room before my mom interrupted it. "How about the four of us go into town for some breakfast this morning? It'll help take your mind off this little scare, Chloe."

"That sounds like a good idea," said my dad. "What do you think, Chloe?"

I begrudgingly agreed, however, what I really wanted to do was read more of Jessica's journal, but that would have to wait. As my parents went downstairs to get ready, I went and woke up Char for the day.

Despite my initial opposition, having breakfast in town turned out to be a good idea. We went to Denny's, which was my sister's favorite—mine too, I guess, but she was always much more excited whenever we got to go. I thought she was going to lose her mind when my parents agreed to it. Sometimes I miss that about being younger—being able to get excited about the little things. I'm not that much older than Char, and I do still get excited about new books, but it's not how it was at her age.

A couple hours later and we were back at the lake house.

Mom, Dad, and Char decided to take a walk down to the beach. I told them I wanted to finish painting my room—which was partially true. Mainly I wanted to read more of Jessica's journal. I waited twenty minutes or so to make sure they were well onto the beach before I closed my bedroom door and pulled out the journal from the floor compartment.

I sat down onto my bed with the journal and carefully opened it to the page I was previously reading. Perhaps it had been my imagination, but the journal seemed unusually warm this time. I noticed it almost as soon as I picked it up, and now that I had it open, it was definitely warmer than it should be—not uncomfortably so, though. I was just about to read Jessica's next entry when I heard it again—that scratching noise coming from behind the library door. It was different this time, though. I can't quite say how, but it was definitely not the same as last night.

I hesitated at first, hoping it would just go away, but it was quite persistent. I put the journal down and moved closer to the door.

"Hello?" I whispered. "Is someone in there?" Silence—then I heard it.

"Hurry," said a quiet, raspy voice. "Come inside, quickly. There isn't much time."

My blood ran cold. *Did I really just hear that? What is happening?* I found myself frozen in front of the door, utterly terrified of what to do next. Before I could make a decision, I heard the voice again.

"Please, you must hurry. Open the door."

"Wh—who are you?" I finally managed to say in response. "Why are you in my room?"

An uncomfortably long silence followed. "There's much you don't know. Please, open the door."

I sighed. After checking that my bedroom door was locked, I went back to the library door and slowly opened it—just enough for me to reach in and turn on the light. To my surprise, I saw an empty room, aside from the books of course.

Uhh…

"Hello?"

"Shh," said the voice. "You'll wake him." Another chill. *Who am I waking?* I thought. *It's just a bunch of books in here, right?* "First book case on the left, third shelf down, red spine."

It didn't take me long to find the one in question. The red spine of this book was quite literally glowing. *This has to be a dream. Maybe I fell back asleep without realizing it?* I reached for the book, and as I took it off the shelf I felt the warmth emanating from it, just like the journal. As I looked at the cover, I realized it was blank—it was just a red book. No dust jacket, no title, no name—nothing. Except it was still glowing, that was something. I stood in the middle of the library, holding the glowing book and wondering what the voice wanted from me.

"Okay," I whispered, "what now?"

"Not here," said the voice, which I now realized was coming from this glowing red book. I don't know what kept me from dropping the book right there and running out of the room screaming, but I managed to keep my composure. "Go back into your room and close the library door as quietly as you can. We must not wake him." There was that sentence again—not wanting to wake him.

A moment later, I quietly closed the library door behind me and sat on my bed with the glowing, talking red book.

LILY

PRESENT DAY

My name is Dr. Lily Burke. I've been a licensed Psychologist for nearly thirty years. I specialize in adolescents and young adults who have experienced trauma at an early age. I imagine this is why the police contacted me on that fateful day in July of 2010. They were brief on the call, which isn't unusual, but what they did say was a sixteen year old girl was going to need some extensive therapy. The officer himself seemed shaken up, and looking back I can understand why. I told them I'd be glad to take on her case.

It was the next day that I met Chloe Henderson. I had offered to see her the night before, but the officer said she was in shock and was being taken to the hospital to ensure she was medically sound. Our first few sessions were nothing but a few pleasantries and long bouts of silence. I knew right away the girl had been through something terrible, but at the moment she was heavily guarding her mind. This is normal.

I remember during our third week of sessions I got her to talk about her family to me, albeit only in minimal detail. Her dad worked for a local paper, the Inforum, and her mom stayed at home. She told me she loved to read and thought maybe she'd be an author one day. I tried to sneak in a question about the lake house, but it was too soon. She shut down almost immediately, and the rest of the session was spent in silence.

I continued to see Chloe on and off for the next six months. We made minimal progress, but there was progress nonetheless. She began telling me more about her life—more of what she enjoyed doing, her relationship with her sister, and whatever else happened to be on her mind that day. It was around month seven when I decided it would be beneficial for her to take a break from seeing me. It would give her a chance to just try and be a kid while she could. I made arrangements with her foster parents to try again in another six months.

CHAPTER FIVE

1

As I continued to sit on my bed with the glowing and talking book, I started to wonder if I was hallucinating. I'd read several books about people who created entirely separate lives in their head, only to realize at the end that it was all fake—what a trip that would be. After another moment of sitting in silence, the movement of the book in my hands brought me back into reality. I closed my eyes and opened them several times, only to find the same glowing book in front of me.

I was not hallucinating.

"How is this possible?" It was the only question I could think to ask at the time. What do you ask a talking book anyway? The book responded shortly after, further solidifying that I was indeed not hallucinating.

"That is a long story," said the book. "I will tell you, but I must ask you a question first. Did you find the journal? Jessica's journal. You did, didn't you?"

It knows about the journal? Did Jessica know about this book? "Yeah, I found it a couple days ago while painting. Why?"

"Thank goodness. I was worried that after what happened before it was taken."

"What happened before? You're being weird." I realized after I said this, that I was the one talking to a book. "Sorry, I just meant, you're not giving me much information, and I'm a little creeped out."

"I apologize. I will do my best to explain. That journal is immensely important. It's the key to you and your family's safety, I'm afraid. So, please keep it hidden whenever you're not using it."

I wasn't sure if I had heard it right. *My family's safety? What's going on here?* I didn't like this conversation, and I had started to regret ever taking this book from the library.

"What do you mean about my family's safety? Is someone going to come for the journal or something?"

There was a brief silence before the book responded. "Let me start from the beginning and explain. This will hopefully answer any questions you have, although I fear it will most likely create more." I said okay and listened to the book's story.

"My name is Dalron. I am an ancient demon from a distant world, bound to the pages of this book. There are six of us in total, all of whom are housed in the library behind that wall. The site on which this structure was built contains a small but not insignificant tear in your universe."

I had to interrupt him. My brain was spinning, and I just couldn't bring myself to accept what I was hearing. "Is this some kind of joke? Please tell me it is."

"I do not joke," said Dalron. "This is real, and there is much more to be told."

Something else came to my mind at that moment, and I didn't mean to ask it, but it just came out. "Are you planning to kill me?"

"No, but there are others who are. If I may, there is more to the story you need to hear."

"O—okay," I managed to say. "Sure."

Before the book could continue, I heard the door downstairs open. I gasped, dropping the book onto my bed.

"My parents and sister are back," I whispered. "They can't find out about this."

"We will discuss more later." The book then went silent as a sense of unnerving washed over me.

I stayed seated on my bed, my eyes locked on the book, unable to turn away from it. *Is this really happening? No, surely it was some complex hallucination. My mom always said I have an overactive imagination. Yes, that must be it.*

A knock on my door pulled me out of my reverie. "Chloe? Can I come in?"

I smiled as the sound of my sister's voice brought me back to reality. I got up and opened the door, my gaze falling not on Char, but on the pile of rocks she was using both arms to hold.

"A successful walk I see. Want to show me?"

She nodded emphatically. "You should have seen it, Chloe. There were sooo many rocks. We walked farther down the lake and I hit the jackpot." She let out a sigh of relief as she dumped all the rocks onto my floor. We both sat down, knees touching, the pile of rocks in between us. *I'll take looking at Char's new rocks over thinking about the demon book any day.*

Char spent the next thirty minutes going over every rock she found. She made sure I saw all the smooth surfaces, the neat indentations here and there, and some that were uniquely colored. I was enjoying the distraction, but it didn't last. I heard the faint scratching against the library door again.

Scratch scratch, scratch scratch.

I felt a tingle go down my spine. *It's not real. It's not real.*

"Did you hear that?" asked Char. *Oh god, she heard it too?*

I shook my head. "I didn't hear anything." I helped pick up her rocks from the floor. "Come on, let's get these into your collection and go downstairs. I want to talk to mom."

2
Therapy Session #253

As usual, I arrived at Lily's office with a hot cup of tea waiting for me. She greeted me with a warm smile and we sat down in front of the fireplace to begin our session. I was still a bit uncertain of where things would go today, especially after everything I dropped on her during our last session. I also couldn't help remembering that I had lied to her about the journal. I still had it, even if I didn't know why.

"Good morning, Chloe," she said. "How are things today?"

I hesitated, but only for a moment. "I'm doing okay. If I'm being honest, I almost canceled this session."

Lily nodded. "Because of how much we uncovered last time." Not a question.

It was my turn to nod. "I—I know it's good to get these things out, but—"

"It's okay," she interrupted. "I think we'll go a different route today." She took a sip of her tea before continuing. "Let's talk about whatever you'd like to talk about. Even if it's as mundane as the weather, or what you did over the weekend, as long as it's light-hearted."

I liked the sound of that, especially after last time. The memories I'd uncovered during our last session, particularly my first encounter with the demon Dalron, were ones I wish I could forget entirely.

So, we spent the session having small-talk. We did in fact briefly discuss the weather before moving on to my dating life (or lack thereof). I told her I had gone on a few coffee dates over the years, and even had a handful of short-term flings. However, nothing had ever come of them. Perhaps it was my trauma subconsciously pushing me toward isolation. I shrugged at the thought and moved on from the topic as swiftly as it had started.

"How has work been going for you?" asked Lily, taking my not so subtle hint that I didn't want to discuss my relationships any further.

When I was younger, I had always imagined myself becoming a writer. I loved reading—couldn't get enough of it—and more often than not I would write extensions of the stories I'd read, or I'd change up a few details here and there to make it my own. Writing came naturally to me, and I was so excited to finish high school so I could focus on becoming the next big name in fiction. That is, until the summer at the lake house happened. It changed everything.

These days, I worked as an editor at the newspaper my father used to work for. Not my dream job, but it still allowed me to use my writing skills. And despite all the lingering trauma from the events I'm trying to work through in these therapy sessions, I do still find time to work on personal writing projects—albeit much less than I'd like.

"Work has been fine," I said, taking another sip of my tea. "I'll admit it's been a bit hard to focus on work since I've started writing down all these memories."

Lily jotted something down in her notebook before returning her gaze to me, and I could have sworn I saw her eyes widen ever so briefly in what looked like deep concern. I wanted to comment, but it vanished as quickly as it had appeared.

"It's understandable that these memories coming to the surface are distracting, and I'm sure painful at times. However, it's vital to your successful recovery that you work through them with me, get to the root of the trauma."

I nodded again, unsure of what to say next. I felt uncomfortable during this session, for reasons I'm unable to express.

"So, Chloe, I know I said we'd keep things light today, but would you be up for getting into a bit of what happened immediately after your encounter with this..." She looked down at her notes briefly. "Dalron individual? You mentioned they were communicating with you through a book?"

I felt my heart begin to race at the mention of his name. I forced myself to take another sip of tea, begging my body to calm down. After almost spilling it in the process, I set the cup back down on the side table and shook my head at Lily.

"I—I don't think I'm ready for that yet. Not aloud anyway. Maybe that can be my goal for next time... try to put in writing what happened after that first encounter." I had to choke out the last words.

She jotted something else down and closed her notebook. "Yes, I think that's a great idea. And I apologize for asking it of you today. We'll end this session here. I don't want to push you too far and risk not having you come back."

Relieved, I followed Lily out of the office and began my walk back home. I couldn't help but feel her gaze on my back as I started toward my apartment.

3

Char and I found our mom down in the kitchen, preparing us some lunch. Despite having said I wanted to talk to her, truthfully I just wanted to get out of my room and away from that library—a phrase I never thought would escape my mouth. Mom smiled at the two of us, focusing on me.

"We missed you on our walk, Chloe. I'm sure Char already told you all about it. What have you been up to? Let me guess, got your nose in another book?"

It took all my effort to force a smile as I nodded. "You know me."

I'd always been good at hiding my emotions, and whether or not my mom ever picked up on it, she ignored any concern she might have had at the moment. She put down two plates in front of us—ham and cheese sandwiches with some chips.

"Eat up, girls. I'm going to make myself a sandwich and join you."

Char began to devour hers as I slowly took a few bites, struggling to take my mind off what the book had told me. *The journal is the key to you and your family's safety.* That's what Dalron had said. My mind wandered to the journal, and how I had felt the heat radiating from it. *What is happening?* I knew then that I needed to read more of Jessica's journal.

"Sandwiches okay?" asked Mom, sitting down across from us with her own plate.

Char nodded, her mouth full of said sandwich. "Mhm!"

I turned my focus back to lunch, but as I went to pick up my sandwich, what I saw caused my stomach to turn. I gasped, my hands falling from the grotesque thing now on my plate, instinctively pushing myself away from the table. Where the freshly made sandwich had been, there was a mold covered pile of rotten food, the smell of which almost caused me to vomit right there at the kitchen table.

My sudden outburst caused both Char and my mom to dart their attention to me, their concerned gazes first going to my plate and then to me.

"Chloe? Chloe, what is it dear?" asked my mom, the concern more evident in her voice.

I covered my nose in an attempt to block the horrendous smell as I looked between them and my plate, unsure of what was happening to me.

"Look!" I pointed at my plate. "Don't you see it?"

"I see your half eaten sandwich," said Char, her plate completely empty. "Aren't you hungry?"

"Doesn't it look… rotten and moldy to you?" Char shook her head.

"Chloe, what on earth are you talking about? I made these fresh just a few moments ago… you watched me, in fact."

I glanced between them again, and then back to my plate. I gasped again. The sandwich was as it had been originally—fresh and half eaten. *I'm losing it*, I thought. *I need to be alone.*

I shook my head. "Ne—nevermind. I think I'm just tired. I didn't sleep very well last night. Char, you can have the rest if you want, I'm gonna go lie down."

Before my mom could protest, I got up and quickly walked upstairs to my room and shut the door, letting out a sigh as I slid down to the floor with my knees to my chest. *What was that? It was real, I know it. I saw it clear as day.* After a moment longer, I forced myself up off the floor and onto my bed, immediately greeted by the red book—one of the six demon possessed books occupying my library.

"It is already starting," Dalron said. "I need to continue telling you our story before it's too late."

I jumped at the sudden vocalization from the book. *I'm going to lose it.* "What do you mean before it's too late? What's going on?" Before he could continue, another knock came at my door.

"Chloe?" It was Char. "Ca—can I come in?"

Oh no, Char. She can't find out about this. Would she even believe it?

"Come back in a bit," I whispered to the book. "Please." After a moment, the glowing red book disappeared, I assumed back into the library. I got off my bed, rubbing my hands over my face, and then through my hair, trying to compose myself before opening the door for Char.

"Hi," I said hesitantly, hoping she wouldn't be too concerned. Her somber eyes told me otherwise, however.

"What happened down there, Chloe? You're really freaking me out."

I sighed and moved out of the way so she could come in. We both sat on my bed as I tried to find a way to explain something I didn't even fully understand. My thoughts kept racing back to the journal. I realized Char was staring at me, waiting for me to say something—anything.

"Look, let's just forget it, okay? I think I've just been reading too many horror novels lately… it's made me a little jumpy."

I knew how fake it sounded before I even said it, and Char's tilted head and narrowed glance confirmed my suspicions.

"Why are you lying to me? We never lie to each other, Chloe."

There was sadness in her voice now, and it honestly cut right through me. She was right, of course. In the eight years we'd been sisters, we'd always been completely honest with each other.

Another sigh from me. "I know, I know. I'm sorry. I just don't know what to think. One minute the sandwich was normal, and then the next it looked like the most disgusting thing I've ever seen."

Scratch, scratch.

I jumped at the sound, my eyes darting to the hidden library.

Char's eyes followed. "What was that? And don't lie, I know you heard it."

"I heard it, but I don't honestly know what it was." I hesitated a moment, wondering if I should tell her what I did know. Thankfully, the commotion downstairs of our dad arriving home saved me from having to say any more.

"Char, Chloe, Dad is home," Mom called out.

Char let out a frustrated huff and went to greet Dad. She stopped and turned to me just before leaving my room. "I wanna know what's going on with you..."

4

The moment my door closed, I felt the temperature in the room become significantly colder. The scratching at the door started again, louder this time. I got up from my bed, rubbing my hands together as the temperature continued to fall, and moved toward the library door. *This has to be some weird hallucination. A very real one.*

I was a few inches from the door, and as my hand slid in-between the opening in the wall, the entire room shifted. I was immediately overtaken by an intense feeling of vertigo. I staggered back, trying to steady myself, as I made contact with my bed—only, it wasn't my bed. In its place was a grotesque equivalent of it, covered in a black dripping ooze.

I gasped, reeling back, my hand now covered in the ooze, and fell to the floor. As I struggled to wipe off my hand, I realized my entire room was changed. While I could tell it was still my room, things were anything but normal. The previously carpeted floor was now covered in dirt, the smell of which assaulted my senses without mercy.

My hands instinctively went to cover my nose as I struggled back on my feet, trying to get a grip on what was happening. *This isn't real, Chloe. It's all in your head.* My heart raced as I stumbled through the thick layer of dirt that was now my floor, trying to get to my bedroom door. I was nearly there when I heard it—a voice so sinister it sent shivers down my spine.

"Hello, Chloe. Come, let's have a chat."

I froze, inches from the door as the sound of that voice reverberated through my body. *It's the library, I know it.* I should have thrown open the door and gotten out, but I couldn't help myself. I forced my legs to move, turning towards the library as I stumbled over the dirt floor.

"That's it," said the voice, "nearly there now."

"Chloe, no!" Another voice rang out. It was familiar, but in the heat of everything, I couldn't place it.

Before I could make it to the door, the room shifted again, vertigo returning in full force. *No, no. Not again.* I braced myself as best I could while the dirt and ooze began to sink into the floor, my room slowly returning to normal. I remained frozen, my mind a haze of dread and disbelief. I managed to get myself back to my bed, which was thankfully ooze-free.

What is going on?

I didn't have much time to contemplate, however, as a moment later I heard footsteps coming up the stairs, followed by a knock on my door.

"Chloe?" It was Dad. "Can I come in?"

I quickly try to fix up my hair as best I can, trying to make it look like I didn't just witness a glimpse of some form of hell.

"Ye—yeah, the door is open."

He steps inside, smiling as he meets my gaze. "Hey there, you okay? Mom told me what happened at lunch. Anything you want to talk about?"

Of course she told him, probably almost as soon as he walked inside.

I sigh. "It was nothing, Dad, really. I think I'm just feeling a little homesick. I haven't been sleeping well." *So many lies, Chloe.*

Thankfully, at least in this case, my dad didn't know me nearly as well as Char. He was a good enough parent, but at the end of the day he was just too wrapped up in his work to ever be that close to me.

He nodded in understanding. "I'm sorry, Chloe. Anything I can do?" His gaze went to the library door, which is open just a hair. "Hey, how about we take a trip to a bookstore or two tomorrow. I'm sure there's plenty of room in that library for some more books."

My eyes lit up. "Really? Yeah, that sounds great!" Despite not being the most present parent, when he was, my dad knew how to cheer me up. At the mention of a bookstore trip, my mind let me forget—however temporarily—about the dirt and ooze that was previously covering my room.

"Good, it's a date then." He leaned over and hugged me. "Come on, let's go downstairs. I'm sure Mom wants to make sure you're okay. How about some tea?"

I nodded. "Sounds good. I'm just going to use the bathroom real quick and I'll be down."

After he left, I got up and went to the bathroom. Standing in front of the mirror, I looked normal enough, but nothing about this situation felt normal. I did what I could to shake it off. As I reached for the faucet to throw some water on my face, I was struck with dread.

Under one of my nails, just enough to see up close, was what looked like dirt.

5

The next day, as I was getting ready, my dad appeared at my door, his slumped shoulders giving away the news before he even spoke.

I sighed. "You can't go, can you?"

"I'm sorry, Chloe. My boss wants the first draft of my story on this place by Monday, and I just don't have the time to spare now."

I should have known not to get my hopes up, but when it came to books, I couldn't help it. Bookstores were my happy place. He always meant well—had good intentions—just didn't often follow through.

Maybe Mom could take me. Char, too. Might help distract both of us.

"Do you think Mom could take me instead? Maybe Char as well?"

He snapped his fingers. "There's an idea. I think the time out of the house would do all of you some good, and give me some quiet to work on this draft. Come on, let's go ask her."

As I moved to follow him out, a brief smell of the death-filled dirt struck my senses, jolting me back to the horrors from moments ago.

I reeled back, my heart beating out of my chest. But the thought of the bookstore surged forward again, drowning out the dread. I caught up to my dad and followed him downstairs.

We found both my mom and Char in the kitchen. Char was doing her best to help with breakfast—mostly by getting in the way. The end result, as usual, was Mom shooing her out so she could focus. She was lax about most things in life, the exception being her cooking. If she was in the kitchen, both Dad and I knew to keep our distance for fear of being reprimanded. Char, however, was still young enough to try and help.

"Hi sweetie," said Mom. "Are you feeling better? You gave me quite a scare yesterday."

I nodded. "Yeah, I'm alright. Sorry about that. Just feeling a bit homesick I think."

My dad put his arm around me. "I had told Chloe I'd take her to a few bookstores today, but work put a tight deadline on me for Monday. Do you think you could take her instead?"

Char's ears perked up at the mention of a bookstore. While she wasn't at my level of fascination with books, being my sister had given her the inclination for reading, by proxy if nothing else.

"Oh, can I come too? Please, Mom. Chloe, pleeeease!"

I couldn't help but smile at her genuine excitement. "It's not up to me, Char. Mom has to drive us. But I was going to ask if you wanted to go either way."

She started to jump up and down at this.

Mom plated some freshly scrambled eggs as some toast popped up out of the toaster. "Sure, I can take the girls after breakfast. It'll give you some quiet time for work, and I can catch up on some crocheting." She took out four plates and handed them to Char. "Go put these on the table, please. Breakfast is ready. Let's eat up so we can go look at some books."

The four of us ate a delicious breakfast of scrambled eggs, hash browns, and toast. It was a staple for us, and one of my favorite meals. And with the confirmation of the bookstore trip, my mind was letting the memory of recent events slip further and further away. I was thankful for the reprieve, however brief it may be.

CHAPTER SIX

1
Therapy Session #254

As I walked with Lily back to her office, I felt the dread growing in my stomach. This was supposed to be the session where I begin to discuss more of the darker happenings during that summer at the lake house. Deep down, I knew I needed to talk about those things—I just wasn't sure if I could.

Lily handed me a cup of hot tea, as was the custom for every session, and sat down in the chair opposite mine. "So, how are you doing today, Chloe? I see you brought some more pages."

I took a sip of the tea, grateful for its comforting warmth. "I'm doing alright." My hand shook slightly as I passed her the pages—each word still buzzing in my head. "I did my best to put into words what happened that day after discovering Dalron. It—it wasn't easy."

Lily nodded, taking the pages and starting to read them. As I waited, my anxiety of what she'd say spiked. *Would she think I'm crazy? Or worse—would she believe me?*

A few moments later, she looked up from the pages, her gentle gaze meeting mine. "Oh Chloe, this is really quite good." She jotted down a few notes as she read back over a few sections. "So, what you've described when your room 'shifted' is very dark. Maybe we should start there."

I shivered at the mention of the room shifting. As I was writing about it, the memories had come back in full force, nearly bringing on a panic attack. Thankfully, I was able to write in short bursts, avoiding remembering too much at one time. I took another sip of tea before responding.

"If I think back to that day, I can still remember the smell. It was awful, Lily. And the image of my bed, or whatever that ooze-covered thing in place of my bed was—I don't think I'll ever truly forget it."

As I spoke, Lily jotted down several things and occasionally looked up at me. This was certainly the most I'd opened up so far in therapy. I should have felt good about that, but there was something gnawing at the back of my mind—it was saying to stop.

"This is good," said Lily, after writing what seemed like a novel of her own. "Very good. I think getting these memories written down is going to help you finally heal." She took another sip of tea before continuing. "What about your relationship with Char? In your recent writing, you mentioned having to lie to her for the first time. How were you feeling at that moment?"

I knew she was going to ask this question, but hearing it out loud still caused the dread in my stomach to crank up several notches. Char was my little sister—I was supposed to be who she looked up to—who she trusted.

I sighed, shaking my head. I felt the tears coming, and I did my best to fight them off, but it was no use. "I—I can't talk about this right now, Lily. I'm sorry."

Lily nodded, closing her notebook as she stood up. "There's no need to apologize. We still made good progress today. Come on, I'll walk you out."

2

My favorite bookstore, From Cover to Cover, opened daily at 10am. Because of this, Mom made sure to plan our drive as such that we arrived just as one of the employees was unlocking the front door. The moment the car was put in park, both Char and I were out of the car, headed for the entrance with our mom bringing up the rear.

As I stepped inside, the unique smell that can only come from a bookstore hit me in a wave of pure bliss. *This is just what I needed.* I took in the rows and rows of books all neatly stacked on their respective shelves. The sound of the baristas getting things set up could be heard from the cafe, which was located at the back of the store, directly in line with the entrance. I always thought this was intentional, to get you to not only buy books, but a drink as well.

I turned to Char. "So, any books you're wanting to get? Or do you just want to look around?"

"Are you really just going to pretend nothing happened with you yesterday?"

The words hit me like a freight train. Any excitement that had come from being in the bookstore evaporated as quickly as it had appeared. *Not now*, I thought. *Please, just let me have this time in my happy place.*

"Please, Char, can we just drop it? I want to look at books with my little sister. Can we do that?"

Her gaze met mine, a look not of anger or frustration, but of sincere concern. After another moment, she finally relented and nodded.

"Great. Should we go over to the young adult first? See if there's anything good?"

She nodded again. "I think the next book in the series I'm reading is out now, so I want to get that."

I smiled, taking her hand as we made our way through the rows and rows of books, all calling out to be brought home with us. We spent the next two hours combing through dozens of books, chatting about this and that (for Char, the conversation almost always found its way to rocks), and trying to decide which books we'd actually buy.

At one point we split up, Char going to look at Fantasy while I found the Horror section. I was looking at the newest release from Stephen King when my mind suddenly whiplashed me back to Jessica's journal. I nearly dropped the book.

Oh my god, the journal!

With everything that had happened over the past twenty four hours, I had totally forgotten about it. Truth be told, I couldn't remember how long it had been since I last read from it. Two days? A week? Time had been weird at the lake house. Days sometimes seemed like hours, while other times an hour felt like it lasted a whole week. *I need to read more of it. Now.*

I quickly picked up the books I had set aside to buy, found Char, and told her we should find Mom so we can get going. To my surprise, she didn't question the sudden urgency of my desire to leave the bookstore. *Does she feel bad for me? Or maybe she just doesn't want to bother with arguing?* I shook the thoughts from my head as we found Mom in the Romance section. No surprise there, she loved that stuff. The sappier the better for her.

She turned to greet us. "Hi, girls. I see you both found an arm-full of books. Ready to go? Maybe we could get some lunch before we go back to the house."

We both nodded in unison, although deep down I wanted nothing more than to be back in my bedroom, reading that journal. *Lunch won't hurt,* I thought, trying to convince myself. *Just go along with the motions and you'll be fine.* The three of us made our way to the checkout. After scanning our dozen or so items, the woman at the counter handed Mom the receipt as Char and I took our separate stacks of books and made our way out the store.

Lunch consisted of a simple sandwich shop we had found that was walking distance from the bookstore. I tried my best to act normal, participating in the various conversations as I munched on a six inch veggie sub I had ordered. Thankfully, the whole meal only lasted about thirty minutes, and then we were on our way back to the lake house—back to Jessica's journal.

3
Lily

As I looked over my notes from the last session with Chloe, I realized things were starting to take a turn. She was making progress, yes. However, we were only just getting to the root of her trauma. The things coming up, I imagine, are really going to determine how the rest of her therapy goes. It's very possible that her brain won't let her accept whatever happened on that summer night at the lake house.

I've had patients struggle to access repressed trauma before, of course. But Chloe's hesitation felt… different. I haven't figured out what made me think that, but whatever it was made me uneasy.

What happened to you, Chloe?

4

I practically ran up to my room, throwing open my door and closing it behind me. My mind was racing. What else would I discover in those long ago written words of Jessica Ambrose? Before retreating to my room, I had lied to my mom, telling her I was really tired and wanted to lie down. She believed it easily enough, but I knew Char hadn't. But, just like at the bookstore, she hadn't pressed the issue further.

Once the door to my room was shut and locked, I quietly pried up the loose floorboard to retrieve Jessica's journal. The worn leather felt soft in my hands—and oddly—warm. Much warmer than such a thing should be. My unrelenting need to see the next entry allowed me to disregard this fact with frightening ease.

I carefully undid the clasp and opened it to the next entry. This one was nearly a month after the one I had last read from June 24th.

July 19th, 1912

Hello, journal,

I'm sorry it's been so long, but things have

been... interesting as of late. I'll do my best to write down what has been happening, but I'm finding it difficult to put some of it into words.

The last entry I made was when daddy started working on my new library. I was so excited. He was going to cut out a space for it, and build a hidden door that was flush with the wall. It didn't take him long to finish, and after I was beyond happy with it. I quickly filled it with my favorite books. Unfortunately, things didn't stay happy for long...

Daddy had said while he was working on my library, he came across a random patch of dirt. He said it smelled awful, like rotting flesh or something. My stomach felt funny at the thought of it. But what was dirt doing inside the walls of my room? Especially on the second floor.

I pried my eyes from the journal entry, my mouth as dry as a desert, my gaze drifting to the library entrance. *The rotten dirt.*

5

Chloe, present day

So, I know what you're thinking. I know because I would be thinking the same thing if I were you. Books possessed by demons, a strange journal written nearly one hundred years ago, and talk of rotten dirt with ooze-covered rooms—sounds crazy, right? Well, I'm here to tell you that it happened—all of it. If you're reading this now, it must mean you're either interested in finding out the truth, or maybe you want to see just how out of my mind I really am.

Whatever your reason for reading (if this even makes it past my hard drive), I hope you'll keep an open mind. What I've written so far has been hard—these memories have been buried in the deepest recesses of my brain for over a decade—having them resurface is unpleasant to say the least. Regardless, my therapist says I need to get this out, so I plan to keep going. If you're still reading at this point, I appreciate you. It's gotten dark in the last chapter, but I warn you—it only gets darker.

6

The mention of the rotten dirt in Jessica's journal entry continued to reverberate through my mind as I sat at the dining table with my family, trying to focus on the dinner conversation. *Did she actually experience the room shifting too? I need to read more.* I had wanted to read the next entry, however, the undeniable link that I seemed to have with Jessica required me to take in these events in smaller chunks. I also wanted to hear the rest of Dalron's story. How had he come to possess that book in my library—Jessica's library.

"How's your food, Chloe?"

My mom's voice brought me out of my reverie, which could have been going on for thirty seconds or five minutes—it was impossible to tell. I started to respond, but as I looked up from my soup, what I saw caused a lump in my throat so big I nearly choked on it.

The dining room floor was covered in dirt. The black

putrid ooze spilled over the table, the pulsing of which made it seem alive. Worst of all, my parents and sister had been replaced by what I can only describe as blobs of the same black ooze.

"No, no, no," I said, shaking my head and scooting back in my chair, nearly toppling over in it. I rubbed my eyes until they ached, desperately wanting to get rid of the awful vision in front of me.

"Chloe? Chloe, what's wrong?" asked my mom.

I forced myself to open my eyes again, and to my relief, the room was back to normal.

I sighed.

"N—nothing, Mom. I'm still just not feeling well. My head is killing me."

I mumbled an excuse and fled the table, avoiding my family's eyes. I quickly scaled the stairs to my room, closing the door behind me. Once inside, I collapsed onto my bed, trying to focus on my breathing to calm down. My heart felt like it was actually going to beat out of my chest.

"Chloe," a raspy voice whispered much too close to me.

I was ripped from my own thoughts, sitting up in my bed, only to find the red book at the foot of it. It pulsed faintly, his voice reverberating through my mind as I tried to get a grip on myself.

"You." It was all I could say.

"Shh," Dalton replied. "You need to keep your voice down. We finally got him back down, but he's getting stronger, and it's becoming more difficult to keep him under control."

My mind raced. *Getting stronger? Who? And why do they have to keep him under control?* I had so many questions.

"I don't understand," I whispered, doing my best to keep my voice down, even though inside I was freaking out a little.

"What is going on? Are you the one giving me these awful hallucinations?"

The silence stretched—thick and breathless, like the air itself was waiting for his response.

"They aren't hallucinations. It's happening quicker than last time… we can't delay anymore. I need to finish telling you our story."

CHAPTER SEVEN

1

At the mention of hearing the rest of Dalron's story, my mind recalled the first book Char had found in her room. *A Night at the Lake.*

"Dalron, my sister found a book in her room a few days ago. It was called—"

"A Night at the Lake," he interrupted. "Yes, I know. And one of the pages cut you, yes?"

I felt my mouth run dry again. I tried, to no avail, to conjure up any amount of saliva. *How did he know?*

"Y—Yes," I managed to say, recalling the surprisingly painful cut the book had given me. Things clicked in my brain a moment later. "It's one of you, isn't it? Or, he? They?"

The already glowing red book pulsed with increased intensity—almost as if in response to the question. *That's a yes if I've ever seen it.*

"That would be Damrok. You and your sister are very lucky he was still quite weak when you found him. That nasty cut you received, while troubling, could have been much worse."

"Why did he cut me?" I had more questions—so many more—but it was the only one I could vocalize.

Another brief pulse.

"You're getting ahead of me, but the cut was a mark. You need to keep that journal safe, even more so than before. You still have it, yes?"

I knew going in that this conversation likely wouldn't make me feel any better, but I didn't realize just how much further it would increase my dread. *Just what did I get myself into?*

"It's under the loose floorboard next to my bed. Why? What's so important about Jessica's journal?"

The book actually grumbled at me. "Keep it on your person, if you can. I don't like it being under the floorboard... Damrok knows of that hiding spot. If you can't always keep it with you, find a new spot to hide it. And don't tell your family about it."

I shook my head in frustration. "Ugh, I don't understand. Just tell me what's going on, please." I didn't like the desperation in my voice, but it was a true reflection of how I felt.

"I have been trying, but you keep asking questions and getting ahead of me."

I deserved that. However, I didn't like the feeling of being scolded by a book. "Okay, okay... I'm sorry. Tell your story."

As I sat on my bed, awaiting Dalron's story, I couldn't help feeling a mixture of emotions. On one hand, I wanted more than anything to know what was going on with the library of possessed books. On the other hand, I just wanted to turn a blind eye to this whole situation and get back to my normal life.

Regardless of what I actually wanted, the story was going

to have to wait a little longer. The pulsing of the book intensified—I could actually feel it in my chest.

"Wha—what's going on?"

"Shh! He's waking. I must go—"

A loud crack came from the library door, startling me half to death. *Oh god, now what?*

Another strong pulse.

"Whatever you do, don't go near the library. Not until he's back under our control."

The book vanished before I could say anything in response—leaving me on my bed in silence.

2

July 20th, 1912

Hello again, Journal,

There are six of them.

Sorry, I should write more than that. I'm scared, journal. I heard it for the first time this morning—the raspy voice coming from my new library. I thought maybe I was dreaming as I'd just woken up. But after rubbing my eyes and confirming I wasn't still sleeping, I heard it again.

It said, "Hello, Jessica."

That was it at first—just kept saying hi to me. I really wish it would have stopped there, journal. It—it showed me things. Awful, terrible things.

I'm sorry, I can't right now. I'll write it down later, I promise.

Bye for now.

3

I looked up from the current journal, my focus again going to the library. *It's all happening again, just like it did to Jessica all those years ago. But what did the book show her? And who was it? Dalron? Or, Damrok?* I had so many questions running through my mind—so many questions that I was still unsure I wanted the answers to.

I went to flip the page to the next entry when a knock came at my door.

"Chloe? Can I come in, please?" It was Char.

I quickly rolled off my bed, kneeled down and concealed the journal under the floorboard, and then got back onto my bed with a random book. "Come in, Char."

My door slowly opened as Char shuffled in, her hands behind her back. She hesitantly made eye contact with me before her gaze darted to the floor. It was obvious she had something on her mind, and I already felt bad about how I'd been acting, so I patted the spot next to me on the bed.

She sighed and joined me on the bed. "I'm sorry for—"

"No, don't." I put my arm around her. "You have nothing to be sorry about. I'm the one who should be apologizing. I've not been myself lately, and I shouldn't have taken it out on you, and worse still, I lied to you about it."

Her eyes teared up as she threw her arms around me, hugging me tightly. "I don't ever want us to fight, Chloe. I love you. You know you can talk to me about anything, right?" She pulled back from the hug, her tear-filled eyes locked onto mine.

I nodded, fighting back tears of my own. "I know, Char. And I love you too." An idea came to me then. "Say, how about a walk on the beach? Maybe we can find some more rocks for you and have a chat, sister to sister."

Her eyes widened, and her mouth turned up in a most genuine smile. "Oh Chloe, really? I'd love that!" She sniffled as she wiped away a few tears with her hand. "Can we go right now?"

"Sure, just let me change real quick and we'll head out."

4

Ten minutes later we were walking through the backyard, passing between the break in the trees, and making our way to the beach. While it was technically a beach, it wasn't one you'd want to sprawl out on and read a book. The sand was a gross dirt color, and since this was an isolated area, driftwood and random trash from the lake lined most of it. Nevertheless, it had rocks, and that was all Char needed.

"How are you liking things here?" I asked, watching her inspect the most recent rock she found. We'd barely left our backyard, and already she'd collected at least a dozen.

It took her a moment, but she managed to split her attention between the rock and me. "It's alright. I get bored sometimes, though. Do you think we'll stay here forever?"

I sure hope not, I thought. It was a little disorienting to think about how drastically my opinion of this place had changed since we first got here. I was initially enamored with its isolation and beauty—but with the journal and the whole possessed book situation, I can't help feeling apprehensive about things.

"I'm not sure, Char. Mom and Dad seem to enjoy it here, but I think once Dad finishes writing his story we'll move back into town."

"Good."

We continued walking along the beach. It was a perfect Summer morning—sunny with just a breath of a breeze to keep the heat under control. I let my mind wander as Char ran ahead of me, most likely toward a super special rock destined for her collection.

How did they, whoever 'they' are, come to possess those books? It was the question at the forefront of my mind. I knew what I really needed to do was continue reading Jessica's journal, and hear Dalron's story. However, it seemed like fate was preventing me from doing either.

The sudden smell of rotten flesh infiltrated my nose, ripping me from my thoughts. I looked down at the beach to find the sand replaced with dirt as black as death. I stood frozen in place, my heart racing as the awful smell intensified.

No, no… not again. Not here. My thoughts flipped to Char. *Where was she?*

I frantically looked around, but all I saw was dirt. I turned towards the lake, hoping she hadn't ended up there. My eyes didn't immediately comprehend what I was seeing—but it didn't take long.

Oh god, are those…

Bodies. The lake had turned blood red, and it was filled with rotting corpses. I shrieked, tripping over my own feet and falling onto the dirt. I began crab walking backwards, wanting to get as much distance between the lake and me as I could. I was nearly to the tree line when I stopped to catch my breath.

You're losing it, Chloe. Get it together.

I felt something clamp tightly around my wrist, once again jolting me back to reality. "What the—" My eyes bolted to the source—it was a corpse's hand coming up from the rotting dirt.

"Chlooooooe" A raspy voice called out as the pressure on my wrist intensified. "You'll join us soon enough, Chloe."

"No!" I clawed at the hand, the rotting flesh peeling off under my nails. I thought I was going to vomit. "Get off me!"

I managed to free myself from its grip as I got to my feet and started sprinting back towards the house. I made it halfway before something caught my leg, and I fell face first into the rotten death dirt. The last I remember before my world went black was hearing (I assumed at the time) Char screaming my name.

5

I woke up to a steady beeping. The first thing I noticed was just how dry my mouth was. *Christ, it feels like I haven't drank in a week.* I went to lift my arm and was met with resistance. Looking down, I saw an IV had been placed in my arm, and on my finger was (I was later told) an oximeter.

A hospital—I was in a hospital room.

I slowly surveyed my room, which had the usual accommodations—stark white walls, two random pictures of dated art, a closet tucked in the corner for clothes and personal effects, and of course the large hospital bed, which I was currently occupying.

What happened to me? If I had had a lot of questions before, the number was now significantly higher. As I began to wake up more, I realized how sore my whole body was. Before I had much more of a chance to consider anything, a soft knock on my door was followed by who I assumed was my nurse. The scrubs and the 'R.N.' on her badge were the main giveaways.

"Oh, you're awake!" She took a quick glance at my vitals and entered them into the computer on wheels that was next to my bed. "How are you feeling? Any pain? Dizziness?"

"Honestly, most of my body is sore, but my head is the worst. What happened to me?"

The nurse typed a few more things into the computer. "I'll get you some meds shortly for the pain. Do you remember how you got here?"

I shook my head. "The last thing I remember—" As I took myself back to the events on the beach, my mind immediately went to Char. "Oh god, my sister—is she okay?" I felt a lump forming in my throat at the thought of something happening to her. My concern was further showcased as the machine monitoring my heart rate began beeping frantically.

"Hey, hey," said the nurse. "Relax, hun. Your sister is okay. Her and your parents have been checking in on you constantly. They went to try and eat something about an hour ago, so they should be back soon."

The beeping began to return to a reasonable rate as I let out a sigh of relief. "How long have I been out?"

Now it was her turn to sigh, although not one of relief. "You were admitted a week ago. You've been in a coma ever since. Your doctor wasn't sure you'd ever come out of it."

A week. A coma. Her words echoed in my brain, bouncing around like they were in a pinball machine.

"Chlooooe," a familiar raspy voice called out.

I turned my attention back to the nurse, but to my horror she'd been replaced by an ooze-covered thing. *Oh god, oh god…*

I frantically blinked a few times and rubbed my eyes hard enough to see stars. *It's not real—you're not real.*

"Chloe?" A normal voice now. I managed to pry my eyes back open, and the nurse was back. "Girl, you gave me quite a scare just now. What's going on?"

I slowly shook my head as the awful image of that pulsing ooze replayed itself over and over again in my mind. "You wouldn't believe me if I told you."

6
Lily

The next few sessions I had with Chloe—255 and 256—have proved to be some of the most informative and concerning. She spent a good portion of session #255 discussing Jessica's journal. I have repeatedly asked her if she still has it, to which her response is always 'no'. I'm not sure why, but I haven't pushed her on it—not yet.

Chloe also tends to focus on the two entities—Dalron and Damrok. The latter being the apparent 'bad' entity that Dalron and the others are trying to 'keep control of'. Toward the end of session #255 I flat out asked her if she thought these were figments of her imagination. After all, she was an avid reader. It certainly wasn't out of the question for this to be true.

She ended that session without any response.

Session #256 was focused on the events of the beach. I must admit, the visual of a blood red lake filled with rotting corpses caught me off guard. At this point, I'm still not sure what really happened to Chloe that day, but I'm inclined to believe it was psychological as opposed to anything in the realm of demonic possession. Still, I can't help but acknowledge her unfaltering commitment to what her mind remembers happened. To her, it was most certainly very real.

7
Chloe

Hello again, reader. I'd like to take a quick break from the past and talk to you a bit more from the present. I just finished another session with Lily an hour ago, so I'm doing some writing to decompress. We discussed the events on the beach today. I told her I didn't remember anything after I tripped and fell face first into the dirt—and that's the truth. My next memory is waking up in that hospital room, apparently from a week-long coma.

In the thirteen years since that happened, I've tried countless times (to no avail) to recall that missing week. It's as if my brain has simply purged it from my memory, never to be seen or heard from again. And maybe that's for the best, all things considered. Or maybe not, I'm not sure. What I do know is that to this day, I wish we'd never set foot in that lake house. But, there's no sense in me going on about that to you. I of all people know you can't change the past—what I wouldn't give to do so, though.

Anyway, I'm going to get myself some lunch and try to write a little more. I think I just needed to write a few paragraphs here as a sort of palette cleanser. Talking about certain parts of my time at the lake house are, as you can surely imagine, quite difficult. So, thanks for letting me vent a bit. I'll see you in the past.

8

As the nurse had guessed, my sister and parents came back fifteen minutes later. It was Char I saw first. She yelled my name and came running, her arms wrapping around me despite the multiple lines in the way.

"Oh, Chloe," she said into my chest. "I thought I lost you." As she pulled back, I saw the tears welling up in her eyes.

"I thought the same, Char." I reached up and wiped away a tear that had escaped her eye. "I'm sorry for whatever happened. I—I don't honestly remember how I got here."

My parents, who had given me a moment with Char, came to the opposite side of the bed, both of their faces marked with worry. My mom was the worst of all, though. She had dark circles around her eyes, and her hair was pulled back into a messy bun. *I don't think she's slept all week.*

"Oh sweetie," she said. "We've been worried sick about you." She carefully leaned down and hugged me. "How are you feeling?"

I shrugged. "I'm okay. My head hurts pretty bad, but I'm mostly just confused."

"That's normal," said the nurse. "I've had your doctor paged to let him know you're awake. He should be in shortly. Being in a coma for any amount of time can cause confusion as well as memory gaps."

"Will they come back? I hate not remembering what happened to me."

The nurse finished up her report and logged out of the computer before turning to me. "It varies with each person, I'm afraid. I'll give you time with your family. The doctor should be in soon."

After the door closed behind the nurse, I turned to Char. "Wh—what happened on the beach? Can you tell me?" She fidgeted with the rock she was holding—a classic tell for her being nervous. "I'm sorry, I'm sure you were scared. You don't have to tell me right now."

She shook her head. "No, I can… I just, there isn't much to tell. I was running ahead, looking for more rocks and then I heard you scream. I'd never heard anyone scream like that before. I turned around and saw you on the beach, face down." She let out a heavy sigh—I could tell she was fighting back more tears. "I thought you were dead. I tried shaking you, but you didn't respond, so I ran as fast as I could back to the house to get Mom."

"And then you brought me here." She nodded. I reached out and took her hand, squeezing it tightly. "I'm sorry again, Char."

"You really don't remember anything?" asked my dad.

I shook my head, also trying not to cry. "Char and I went for a walk. I wanted to clear my head, and I figured she could find some more rocks. Everything was fine until—"

Scratch, scratch.

The familiar sound interrupted my thoughts. It sounded like it was right next to me—no, closer. I turned to my left in mid-sentence, only to see nothing. "Sorry, I thought I—"

I turned back to face my parents, but what I was met with froze the blood in my veins. A dark malevolent figure hovered above me, inches from my face. I tried to scream but no sound came out. The temperature in the room plummeted as the figure began to laugh.

"Oh, Chloe. Nobody can help you. You're already too far gone." Its face, grotesque and dripping with ooze, shifted to form a sinister smile. "Time to sleep." The figure rushed towards me and then vanished. And for the second time in a week, my world went dark.

CHAPTER EIGHT

1
Therapy Session #257

If you're still here, I'm sure this won't come as a surprise, but I almost canceled my next therapy session with Lily. I knew going into it that she would want to discuss the events at the hospital, as well as the ones that followed. She still swears that getting all of this written down will help—I'm yet to be convinced.

"How are you feeling today, Chloe?" As usual, Lily was seated in a chair across from me, our cups of tea in between us. The crackling of her fireplace put me at ease as it always did.

I took a sip, cherishing the warmth of the tea as it went down my throat. "I'm doing alright, I think." I hated the uncertainty in my tone, and of course, Lily latched onto it.

"Tell me about that. What's on your mind?"

I hated that question.

"I just don't know about this exercise we've been doing…

writing down all these memories I've been suppressing. I know it's not good to do that, but I'm starting to wonder if I need a break."

"I see." She jotted a few things in her notebook before looking back at me. "Chloe, I really think you should keep going… you're very close to breaking through this trauma. I'd hate to see you stop now and potentially lose the progress you've made so far,"

I sighed and took another sip of my tea—Earl Grey today. Lily was right, of course. I guess that's why she was the therapist, and I the client. Deep down, I think I did want to stop suppressing the trauma from my time at the lake house. And if continuing on with the writing exercise was how to do that, I needed to keep at it.

"Okay, okay," I said. "You're right. Can we skip our next session at least? I think the coming events will take me a bit longer to put into words."

She smiled, closing her notebook. "That sounds like a very reasonable compromise. Come on, I'll walk you out."

2

I woke up again, but this time I was back in my room at the lake house. My body was as sore as ever, and my head felt like it was going to explode. I looked around, still processing everything, wondering how long I was out this time. *How did I get back here? Am I losing it?*

"Chloe?" that familiar voice whispered. Before I could respond, the glowing red book appeared at the foot of my bed. It pulsed with a strength I felt to my core.

"What is going on?" I asked, doing my best to keep my voice down. "I feel like I'm losing my mind." The statement was made even more ridiculous by the fact I was saying it to a talking book, possessed by an ancient demon.

"The journal… do you still have it??"

At first my brain didn't process the question. After being in a coma for a week, followed by waking up back in the lake house with no memory of getting there, I was a bit out of it.

"Chloe!" Dalrun whispered again, more urgently, pulling me back to the moment.

"Sorry, uh… it should still be down here." I rolled out of bed, much to the dissatisfaction of my body, and pulled the journal from underneath the loose floorboard. "Here it is."

Dalrun sighed. Yes, the demon book actually sighed at me.

"I thought I told you to find a new hiding spot for that. This is serious, Chloe. If you were at the hospital much longer, it might have been too late for me to save you."

I heard Dalrun's words, but again my brain refused to comprehend what was actually said. *Too late to save me?* The number of questions I had was growing by the second, and each time it did, my headache grew with it.

"Save me from what? What happened to me?" I put my face in my hands in frustration. "I don't even remember how I got back here."

"I am sorry, Chloe. I wanted to tell you more of the story earlier, but Damrok has been… complicating things. He's grown much stronger and more quickly than last time."

I sighed. "Last time? You mean, Jessica?"

The book pulsed. *That's a yes.*

"Dalrun, what happened to Jessica? You know, don't you?" I asked the question, but a large part of me didn't really want to know the answer. The other part of me already had an idea of the answer.

SMACK!

The deafening crack against the library door sent my heart into overdrive. "Fucking christ, what was that?"

The pulse intensified once more. "He's so very close, Chloe. I must go again. Read Jessica's journal. You mustn't let anything else sidetrack you. I wish I could tell you the rest, but this takes precedence. Read every entry… with haste."

Dalrun vanished, once again leaving me alone with my thoughts.

3
Chloe

Present Chloe here again—taking another break from writing down these traumatic memories. There's one thing I wanted to clarify, which I don't think I've done a good job of so far. For the longest time after 'it' happened, I couldn't remember any of it. I've asked Lily countless times about the night in June, the one where the police found me outside the lake house. She dodges the question, always telling me that I need to remember it for myself.

I won't lie, this was incredibly frustrating at first. Here I was at just sixteen, having experienced some form of intense trauma but with no memory of it, and a therapist who was refusing to fill in any of the (many) blanks I had at the time. This was a big source of contention between the two of us, and one of the main reasons I stopped going to therapy so many times. Looking back, though, I now understand why she did it.

It happened on July 27th, 2010—at least that's what I'm told. To this day I don't remember that night at all. As I've been working with Lily and writing all this down, memories have obviously started coming back. However, it's in fragmented pieces, many of which are quite blurry. The night of, along with the following few weeks, are still a complete black void in my memory. To call it frustrating would be an understatement.

For better or worse, throughout my sessions with Lily I began to remember—filling in the missing pieces of my memory. To this day there are still parts I haven't gotten back, which is why there will be things missing in this journal or whatever you want to call it, but I plan to write down everything that I can. I feel like people need to know what happened.

Whether they believe it is up to them.

4

July 21st, 1912

Hello Journal,

I apologize for my last entry, I was really freaked out. I guess I still am, but I want to try and talk about it now.

A different one spoke to me this morning. It called itself Dalrun. It's this blood red book, no title, author, or anything... but it glows. I could have sworn it pulsed as well, usually while it was telling me its story. I'm going to do my best to relay it here, but there was a lot, so maybe I'll add more in future entries if I remember something I missed in this one.

Dalrun's Story

Like I mentioned in my last entry, there are six of them. They are demons from a very dark and distant world. The lake house was built on what Dalrun calls a weak spot in our universe. Because of this, a tear formed, opening a sort of gateway between our two worlds. I know, journal, it sounds wild... but I promise it's true. I've seen it.

The six demons, who Dalrun says are referred to as Damrok's Deathbringers, stumbled upon the tear and became stuck on our side of it. They aren't able to exist in our universe long-term unless they occupy a vessel. This is how they became attached to six of the books in my library.

Okay, journal, Mom is calling me down for dinner. I'll continue this story real soon, I promise.

I sat in my bed, Jessica's journal open on my lap, as I replayed her words in my head. I struggled to keep my thoughts straight. This was due to both a throbbing headache, as well as the looming sense of dread I couldn't shake. *This is too much. It's bad enough I once again don't remember how I got somewhere, and now these so-called Deathbringers are stuck in the library in my room.*

I needed to keep reading.

July 22nd, 1912

Hi Journal,

Today I'll try to get through the rest of what Dalrun told me. Let's see, where was I? Oh right, they attached themselves to books in my library as they need a vessel to survive in our world.

While they're all quite old compared to humans, apparently Damrok, their leader, is ancient. Dalrun said there isn't even a number in our world large enough to describe his true age. Anyway, Dalrun is his second in command, and the other four more or less serve as henchmen.

After they came through the gateway and attached themselves to the books, the shockwave of their combined energies caused the tear to close—at least mostly. Dalrun said there's still a miniscule tear remaining, but it's much too risky to attempt crossing over at that size. And yes, I did ask him if there was any way to reopen it. He seemed hesitant to answer me.

Oh, sorry journal, Mom wants me to come help her with something downstairs. I'll write more soon.

I turned the page to read Jessica's next entry when I heard the front door open and close. I got off my bed and quickly left my room, making my way downstairs. As I entered the kitchen I ran into my dad—quite literally.

"Chloe!?!" Between his tone, wide-eyed stare, and the blood draining from his face, it was clear I was the last person he expected to see. "What are you doing here?" He took a quick step backward.

Well that's not great.

"Wha—what do you mean, Dad?" It took a solid thirty seconds for him to voice a response.

He shook his head. "You're supposed to still be at the hospital."

The hospital? Wait, Dalrun mentioned something about saving me. Did he—

My dad's hands on my shoulders pulled me from my thoughts. "Yo—you're actually here. What's going on, Chloe?"

"I—I don't know." I felt myself on the verge of tears, which I had to fight with all I had.

He motioned to the dining table. "Come on, I'll make us some tea and we can chat."

I took a seat at the table, trying my best to remember anything at all. "Where's Mom and Char?"

Dad brought two cups and a kettle to the table. "They went out for a walk along the beach. It took some convincing, but your mom finally got Char to go. After what happened to you, I can't say I blame her."

You and me both.

"Why did you think I was supposed to still be at the hospital? When did you last see me?"

He sighed, pouring us two cups of tea from the kettle. "Uh, two days ago I guess? You finally came out of your coma after a week, and you were talking about what you remembered from the beach… and then—" He stopped pouring, frozen in place as if he just remembered a terrible fact.

"What, Dad? Then what?"

A moment later he let out a ragged sigh. "You were talking about the beach, trying to remember more of what happened, and then you just stopped midsentence. Your vitals suddenly spiked… and your expression… Chloe, I don't know if I'll ever get that image out of my head."

"Wha—what did it look like?" I felt a tingle crawl up my spine, thinking back to that horrible figure in the hospital.

His face darkened. "You looked like you'd seen the devil himself."

I stared in silent horror from across the table as I took in what my dad just told me. I vividly remember the grotesque figure I saw in the hospital. I sure wish my brain would choose to forget that particular memory as opposed to how I got back home. But, in reality, I think I knew how I got back home. Dalrun used some demonic spell to save me. That's the only explanation that even remotely makes sense, which should tell you all you need to know about the situation and how messed up it was.

My dad waved his hand in front of me. "Chloe? Please don't do this… not again."

"I'm here, sorry. I—I don't really know what's going on. I just woke up back in my room a few minutes ago. Oh god, Mom is going to freak."

As if the universe heard me, the front door opened, Mom and Char coming in from their walk. "We're back, hun," said Mom, walking through the kitchen and into the dining room. "Should we go see Chlo—"

Her eyes widened just as my dad's had, if not more so. "Ch—Chloe. Wh—what are you doing here?"

It was the second time in only a few minutes I'd been

asked what I was doing in our house. It wasn't a great feeling, to say the least.

Char, on the other hand, just seemed to be happy to see me awake. "Chloe!" She darted to me, wrapping her arms around me. "You're okay?"

I returned her hug. "I think so, Char, yeah." I looked up, meeting my mom's eyes. "But, I really don't know what's going on. One minute I'm awake in the hospital and talking to you guys, the next I wake up back here in my bed."

My mom turned to my dad. "Should we take her back in? I don't like this one bit."

After a moment of silence my dad shrugged, not really having much of an opinion. "I don't know, dear. She seems fine now. I'm not sure the hospital would be able to tell us much more than she has anyway."

I don't recall much of what else was said that day, but ultimately my mom was convinced to just let it go. Yes, I was surprised as well. My mom is usually not the type to do that, especially when it involves one of her daughters. Nevertheless, I knew now more than ever that I needed to finish reading Jessica's journal.

CHAPTER NINE

1

July 23rd, 1912

Hello Journal,

Today was a bad day.

I think I got a glimpse of their world. I'm still not sure if it really happened, or if it was just a terrible dream. I was cleaning up my room when I heard this subtle scratching sound coming from my library. As I approached the door it stopped. That was when it happened—my room sort of... shifted. That's the best way I can think of to describe it. The next thing I knew I was on my floor, only it wasn't my bedroom floor anymore—it was dirt.

The dirt was awful, journal. It had a smell I don't think I'll ever forget. It smelled like... like it was full of dead things. My stomach is upset again from just writing it down.

It got worse though, if you can believe it.

Once I managed to stumble back up to my feet, which were now covered in the death dirt, I noticed the walls and ceiling of my room were—wrong. They were moving. It was like this black goo. I don't really know how else to describe it. I was so scared. I ran as best I could through the dirt and got out of my room as fast as I could.

Thankfully, as soon as I opened the door of my room, everything went back to normal. Ugh, I don't know what to do. Should I tell my parents? They'd never believe me. I guess for now I'll just keep writing it down here, journal. I need to rest, I'm so tired.

Until next time.

2

I once again looked up from the entry I just read. It was as if I was re-living what Jessica went through when she lived here. *Do I really want to read the rest?* I of course knew the answer as soon as the question formed in my mind. Whatever happened to Jessica all those years ago is now happening to me. And based on the urgency in Dalrun's voice, her journal is vital to my (and my family's) safety.

July 24th, 1912

Hello Journal,

Last night was a lot, but I got some rest, and now I feel better. I just finished talking to Dalrun. He told me the rest of his story. I wanted to get it written down right away so I don't forget anything. It's... it's a lot.

I left off with Dalrun avoiding my question about reopening the tear that connects our two worlds. I asked him again this morning. He said there is a way, but I got the feeling he didn't want to expound on his answer. I pushed him though, and he finally gave in.

I kinda wish he hadn't.

Apparently to reopen the tear, a large amount of life force needs to be collected and directed towards the tear. He explained further that whenever they take control of a living host, they're able to suck out the host's life force bit by bit until there's nothing left.

I'm so scared, journal. Dalrun said Damrok is trying to get to me... to take control of my body. He... he wants my family. I finally understand it now. They want to reopen the tear to get back home, but they would need the life force of my entire family to do it. I... I can't let that happen. But, I don't know what to do.

I need to go for now. I hope to write again soon.

I nearly dropped the journal as the sudden realization of what was happening (again) hit me like a freight train. *The mark. Damrok has me. Is it already too late?* As my pulse quickened, I flipped ahead in the journal, and noticed there were only two entries left.

I did my best to calm myself as I read the second to last entry.

July 25th, 1912

Journal!

So much to write, so little time.

Dalrun spoke to me again tonight. He said there wasn't much time left, and that he wanted to protect me and my family. I asked him how. He did some fancy magic, which put a protection spell on my journal. Dalrun said it would protect the holder and anyone they deem worthy. He said to do this, I just need to write their names on the first page of the journal. So, I wrote my family's names, of course.

As I did, the page flashed a bright yellow, almost as if it glowed in acknowledgement. I wouldn't have believed it if I didn't see it myself. Is that really it? Will the spell Dalrun put on this journal really save me and my family?

All I can do is hope, journal. I'm so scared. I've been continuing to hear scratching coming from the library. It must be Damrok.

Please, journal... protect us.

I looked up from the journal, my eyes darting to the library door. *And the scratching, too. It really is all happening again*. I felt conflicted—read the last entry or get a pen. I decided to read the last entry and get it over with.

I took a deep breath and turned the page. I felt the warmth of the journal intensify as I did.

July 26th, 1912

Journal! It worked!

My family is safe.

But, it wasn't easy, and it was so scary. I'll do my best to recap. Damrok came for me tonight. He wanted my life so badly, but thanks to having the journal on me, Dalrun's protection spell managed to overpower him, sending him back to a slumber in the library.

It happened when I was heading downstairs for dinner. I had left the journal on my bed, which looking back, was a stupid thing to do, especially after Dalrun insisted I never leave it out of my sight. Anyway, the moment I opened my door to leave, I felt my world going dark. The smell... that putrid smell of the rotting dirt returned. I don't know how I managed, but I did all I could to ignore the smell and turn back to my bed. I dove onto the bed, grabbing the journal and holding it tight to my chest.

A super bright flash of light flooded my room. The smell began to fade as the library door opened and slammed shut. I stayed on the bed for a few moments, unsure if it was safe to get up. The heat from the journal was intense, but also comforting.

A moment later, Dalrun appeared—the glowing red book was next to me on the bed. He told me the spell on the journal had managed to restrain Damrok. He won't be a threat as long as he remains undisturbed. Dalrun expressed how important it was to always keep the journal nearby, just in case. If someone is reading this, then you found the hiding spot I made for it under a loose floorboard by my bed. I hope you never have to use it, but if you do, know that this journal is special and will help.

Keep it safe!

3
Therapy Session #258

After skipping the previous session, I was more nervous about therapy than usual. However, as soon as I took a seat next to the fireplace with a cup of hot tea in my hands, my nerves began to calm.

"So, how has your last week been, Chloe? Did you get more writing done?"

I nodded, taking a sip of the delicious tea. "My week has been okay. And yes, I did." I pulled out the handful of recently printed pages and handed them to Lily.

As usual, I felt like time stood still while she read the pages. I watched her pause every once in a while and jot something in her notebook. After what seemed like ages, she put the pages down and returned her gaze to me.

"This is still very good, Chloe. I noticed you stopped right before the night of it happening. I assume that was intentional."

I forced a laugh, my discomfort evident. "Y—yeah, I needed to step away from it for a while. I'll be honest, I'm terrified to think about the remaining few hours of my time there."

Lily nodded, jotting another note down. "That's very understandable. We're finally about to broach the core of your trauma. I want to reinforce how important it is for you to see this through. I know you're scared, but trust me, this will ultimately help you."

I sighed. "I know, I know. I've come this far, and even

though it's terrifying to think about, I know I'd regret it if I stopped now." I took another sip of tea, and as I did, an idea came to me. "Lily, would it maybe be okay if I worked on writing this last piece here during our session? Having you here would really help. Your presence always calms me."

"I think that's a wonderful idea." She smiled, closing her notebook. "I'll get us some more tea. You go ahead and make yourself comfortable and get to writing."

4

After I finished reading the final entry in Jessica's journal, I knew I needed to find a pen. Naturally, despite almost always having multiple pens in my room, I couldn't find even one. Frustrated and not thinking, I flung open my door and headed downstairs, leaving the journal on my bed, just like Jessica had done almost one hundred years ago. I no more than hit the first step when my world went black as a faint smell of rotting dirt infiltrated my senses.

5

Therapy Session #258 (cont.)

Scratch, scratch

My pulse quickened as I looked up from my laptop, wondering if I imagined that scratching. *No, it was only in your head. These memories can't hurt you, Chloe.* I frantically looked around the once calming office, desperate for Lily's presence, but she was nowhere to be found. I thought I heard her saying my name, but then it hit me—that smell of rotten dirt.

"L—Lily? Are you ther—"

SO CLOSE.

CHLOE
CHLOOOOE
YOUR… BODY…
IS… MINE.

6

Lily (transcribed from discovered audio note)

At our last session, Chloe had asked if we could skip a week to give her time to collect her thoughts. As much as I wanted to push her, I knew it would ultimately benefit her more to give her that extra time. I'm worried that as we get closer and closer to the root of her trauma, she'll back out from this exercise and never truly come to terms with what happened on that fateful summer night.

Jump to our current session—I just finished reading Chloe's most recent chapter. The details she's remembering are good. She stopped right before the night where everything happened. I could tell she was hesitant to continue, as I expected, but I encouraged her to see this through. She's so very close.

This encouragement seems to have worked, as Chloe asked if she could finish writing the last section during our session. I really wasn't expecting this, but if it helps get her through it, I'm more than willing to let her use our session. They're her time to use for what she wants, after all.

I think something bad has happened. I went to get Chloe and me some more tea, and when I came back, she was staring straight ahead, motionless in the chair. Her face is as white as a ghost, and her eyes—god, I don't think there are words to describe the horror I saw in them.

"Chloe? Chloe, can you hear me?"

"Hello, Dr. Burke. It's so nice to finally meet you." *(a chilling evil laugh erupts from the recording)*

"C—Chloe?"

"Not anymore…" *(a demonic voice replies)*

(Concerning silence fills the recording for a few seconds)

"Oh god, it's you. So, it was all true then…" *(Lily's heavy breathing can be heard as she stumbles backwards)*

"I'm going to enjoy this."

Abrupt and frantic commotion takes over the recording. Lily's screams of anguish are heard as her skin is forcefully ripped from her body. Every bone in her body is violently crushed, bringing a sudden stop to the screams, replaced by a visceral slurping. A moment later, the recording fades to silence.

EPILOGUE

JULY 27TH, 2010

As the officers frantically tried to understand what happened inside the lake house on Alpine Way, Chloe sat motionless in the back of an ambulance, a warm blanket draped over her shoulders and the journal gripped tightly in her hand. She stared ahead unblinking while the paramedics took her vitals, making sure she was okay, at least medically.

"Is that your journal?" asked one of the paramedics, pointing to it.

After a moment, Chloe nodded. She opened it to the first page. The paramedic taking her blood pressure noticed the spike.

"Whoa, easy there. You're safe. What's wrong?" He leaned in and looked at the open page of the journal, but it was blank.

Her hands began to shake as she slowly shook her head. "I didn't write that..."

Centered on the page.

Written in dark red.

Damrok.